CE

13+

Science

Revision Guide

Boost

GALORE PARK

AN HACHETTE UK COMPANY

About the author

Louise Martine has worked in the educational publishing business for over 19 years, during which time she has commissioned and published educational textbooks and revision guides for Years 3 to 11 in mathematics, biology, chemistry, physics, study skills and geography. Louise's publications include Galore Park's *11+ Mathematics Revision Guide, Study Skills for 11+* and Rising Stars' *Achieve Times Tables*.

In her spare time she is a prep-school governor with specific responsibilities for education and has contributed to many articles on how to prepare children for their exams. As a mother to four children she has a wealth of experience and insight into schools, teaching, learning and how to guide the various personalities through their school life, their revision, school exams, CE exams (still one child to go!) GCSEs, A level and university exams.

Photo credits

(From top to bottom then left to right)

p11 © phonlamaiphoto/Adobe Stock, © tonaquatic/Adobe Stock, © mrallen/Adobe Stock, © Emilio Ereza/Adobe Stock, © Africa Studio/Adobe Stock, © Fotofermer/Adobe Stock; p18 © Nevada31, © artempohrebniak, © Public Health England in association with the Welsh Government, Food Standards Scotland and the Food Standards Agency in Northern Ireland; p21 © dennisvdwater/Adobe Stock; p25 © Oxford Media Library/Shutterstock; p31 © frenta/Adobe Stock, © nobeastsofierce/Adobe Stock, © frenta/Adobe Stock, © jovannig/Adobe Stock; p45 © Romolo Tavani/Adobe Stock, © Kimberly Boyles/Adobe Stock, © ToriNim/Adobe Stock, © tom/Adobe Stock, © Елена Петрова/Adobe Stock, © micromonkey/Adobe Stock; p52 © Patrick Foto/Adobe Stock, © Angela Hampton Picture Library/Alamy; p58 © mzphoto11/Adobe Stock, © Natalya Osipova/Shutterstock, © bajita111122/Adobe Stock; p71 © W.Oelen; p77 © Dorling Kindersley/Alamy; p84 © Paul Maguire/Adobe Stock; p101 © divgradcurl/Adobe Stock; p105 © Awe Inspiring Images/Adobe Stock; p110 © petert2/Adobe Stock, © nordroden/Adobe Stock; p111 © o_shi/Adobe Stock, © kanin/Adobe Stock, © ngel.ac/Adobe Stock, © Studio Harmony/Adobe Stock; p112 © Robert/Adobe Stock; p125 © aleksey/Adobe Stock; p132 © Den/Adobe Stock; p133 © Wayhome Studio/Adobe Stock; p134 © petrroudny/Adobe Stock, © G3D Studio/Adobe Stock; p136 © Iakov Kalinin/Adobe Stock, © Brilliant Eye/Adobe Stock, © Ansario/Adobe Stock, © alg2209/Adobe Stock; p137 © Kseniya/Adobe Stock, © Kseniya/Adobe Stock; p138 © Tom Fenske/Adobe Stock; p141 © Sven Bachstroem/Adobe Stock ; p147 © harunyigit/Adobe Stock; p149 © realstock1/Adobe Stock; p153 © Tim/Adobe Stock; p154 © petaran/Adobe Stock, © Markus Mainka/Adobe Stock; p161 © leekris/Adobe Stock, © leekris/Adobe Stock, © MJB/Adobe Stock; p162 © gumao/Adobe Stock, © Christoph/Adobe Stock, © Andrey Armyagov/Adobe Stock, © EurekA_89 Gervasio S

Every effort has been made to trace all copyright holders, but if any have been inadvertently overlooked the publishers will be pleased to make the necessary arrangements at the first opportunity.

Although every effort has been made to ensure that website addresses are correct at time of going to press, Galore Park cannot be held responsible for the content of any website mentioned in this book. It is sometimes possible to find a relocated web page by typing in the address of the home page for a website in the URL window of your browser.

Hachette UK's policy is to use papers that are natural, renewable and recyclable products and made from wood grown in well-managed forests and other controlled sources. The logging and manufacturing processes are expected to conform to the environmental regulations of the country of origin.

Orders: **Teachers** please contact Hachette UK Distribution, Hely Hutchinson Centre, Milton Road, Didcot, Oxfordshire, OX11 7HH. Telephone: (44) 01235 400555.
Email: primary@hachette.co.uk. Lines are open from 9 a.m. to 5 p.m., Monday to Friday

Parents, **Tutors** please call: 020 3122 6405 (Monday to Friday, 9:30 a.m. to 4:30 p.m.). Email: parentenquiries@hachette.co.uk

Visit our website at www.galorepark.co.uk for details of other revision guides for Common Entrance, examination papers and Galore Park publications.

ISBN 978 1 3983 4092 3

© 2022 Hodder & Stoughton Limited
First published in 2022 by Hodder & Stoughton Limited
An Hachette UK Company
Carmelite House
50 Victoria Embankment
London EC4Y 0DZ

www.galorepark.co.uk

Impression number   10 9 8 7 6 5 4 3 2 1

Year   2026 2025 2024 2023 2022

Typeset in India

Printed in Spain

A catalogue record for this title is available from the British Library

Contents

Introduction

This new edition of the *Common Entrance 13+ Science Revision Guide* is intended to assist you in your revision and contains a review of all the 13+ content from the new ISEB specification that will be examined. If you would like more detail, this can be found in *Common Entrance 13+ Science for ISEB CE and KS3* by Ron Pickering. The book covers all three areas of science: biology, chemistry and physics.

Some of the 11+ Science content is reviewed as 'Preliminary knowledge' in this book. However if you feel the need to revise the earlier work, please refer to Galore Park's *11+ Science Revision Guide* by Sue Hunter. This knowledge will be assumed and can be examined.

Preliminary knowledge

These sections summarise the preliminary knowledge acquired from the ISEB 11+ Science specification.

Preliminary practical activity

Reminders are given about the practical activities covered in the ISEB 11+ Science specification. These too may be examined.

Test your preliminary knowledge

Exam-style questions are set to test your preliminary knowledge.

The book reflects the three command words of the learning outcomes of the new ISEB Science specification in terms of content and assessment.

Know Specifies the knowledge to be learnt and recalled.

Understand Requires you to apply your knowledge to familiar and novel situations.

Recognise Requires you to show an awareness of the significance of your knowledge and understanding to science and society.

Each chapter also includes, where relevant:

Recommended practical activities

These activities are regarded by the ISEB as the minimum practical experience pupils should know. They will be assessed and descriptions of methods may be asked for. Further details of these activities and more, can be found in *Common Entrance 13+ Science for ISEB CE and KS3* by Ron Pickering.

Ask yourself

Throughout the revision guide there are questions that prompt recognition and the development of scientific attitudes using the knowledge and understanding gained. This is designed to encourage critical thinking. See the answers section for responses and further discussion opportunities around these.

Exam-style questions

At the end of each chapter, there is a short set of exam-style questions. In addition, there are three 'Test yourself' sections: one each for biology, chemistry and physics. The answers are given at the back of the book.

Good luck with your revision and exams! **– Louise Martine**

Your exam

Pupils sitting ISEB Common Entrance at the end of Year 8 may sit a single combined paper (for Foundation Level) or separate papers in biology, chemistry and physics (for Level 2). There is also a scholarship paper: this is a single paper divided into sections for biology, chemistry and physics. Check with your school which examination you are going to sit.

Within each examination, marks will be allocated according to the following assessment objectives:

A01 remembering and understanding   knowing and understanding

A02 applying and analysing   thinking and working as a scientist (TWAS)

A03 evaluating and creating   recognising

Examination questions will be drawn from all parts of the specification. Candidates will also be expected to think and work as a scientist, analysing and evaluating scientific knowledge, and applying it to unfamiliar situations.

Command words in the 13+ Examination

The examination papers at 11+ and 13+ will use a system of command words which indicate the type of response required and the depth of the answers. The system used is compatible with that used by GCSE awarding bodies in England and Wales.

It is recommend that candidates become familiar with the use of command words in the examination papers. These are useful metacognitive skills which will form a foundation for their future studies in science.

The depth of answers expected is given by the number of marks and the number of lines in a response, and candidates can be guided by this.

The following command words will be used:

Calculate	Use the numbers given in the question to work out the answer. Calculate the mass of air in the laboratory, in kg.
Compare	Describe the similarities and/or differences between things. Compare how the mistletoe plant and the tree obtain their water.
Complete	Add missing information to a table, diagram or graph. Complete the table by calculating the mean time taken to slide down the slide.
Describe	Recall some facts, patterns in results or a sequence of events accurately. Describe the relationship between current and number of paperclips lifted.
Discuss	Write about a topic in detail, taking into account different ideas, and sometimes different opinions. Discuss the advantages and disadvantages to society of these projects.
Draw	To produce, or add to, a diagram or graph. Draw straight lines to match the parts labelled from V to Z with the word which best describes them. Draw a curve of best fit through the points.
Estimate	Give an approximate value to a quantity. Estimate the speed of sound.
Explain	State the reasons for something happening, using scientific knowledge and understanding. Explain how these results suggest that the pondweed is photosynthesising.
Extend	Draw on a diagram or graph to complete it. Extend the light ray to show ...

Give	Write a brief response. Only a short answer is required, not an explanation or a description.
	Give the correct unit.
Identify/Name	Write the recognised name of something.
	Name the parts of the cell.
	Name the force which keeps a satellite in orbit round the Earth.
Label	Add appropriate names to a diagram.
	Label the nucleus on the diagram.
Measure	Read a scale to get a quantity.
	Measure the width of the cell along line X.
Outline	An answer that gives only the most important points. Detailed responses are not needed.
	Outline a difficulty the pupils may experience in obtaining an accurate result.
Plot	Add points to a graph using data given.
	Plot points on the grid below for the height of the dough left at 35 °C.
Predict	Write a plausible outcome.
	Predict the position of an ice cube when it is added to this mixture.
Show	Provide structured evidence to reach a conclusion.
	Show that the volume of the laboratory, in m^3, is about 190 m^3.
	Show clearly on your graph how you obtained your answer.
State	Give a brief definition of facts.
	State the equation which links density, mass and volume.
	State two features of mammals.
Suggest	Apply knowledge and understanding to a new situation.
	Suggest three reasons why Chlorella might be useful as part of a balanced diet.
Use/using	The answer must be based on the information given in the question. In some questions, candidates will use their own knowledge and understanding.
	Using the tables at the beginning of this question, describe all the colour changes ...
Write	Only a short answer is required, not an explanation or a description.
	Write the word equation for aerobic respiration.
	Write a food chain for the pyramid of numbers.

Tips on revising

Get the best out of your brain

- Give your brain plenty of oxygen by exercising. You can revise effectively if you feel fit and well.
- Eat healthy food while you are revising. Your brain works better when you give it good fuel.
- Think positively. Give your brain positive messages so that it will want to study.
- Keep calm. If your brain is stressed it will not operate effectively.
- Take regular breaks during your study time.
- Get enough sleep. Your brain will carry on sorting out what you have revised while you sleep.

Get the most from your revision

- Do not work for hours without a break. Revise for 20–30 minutes, then take a 5-minute break.
- Do good things in your breaks: listen to your favourite music, eat healthy food, drink some water, do some exercise and juggle. Do not read a book, watch TV or play on the computer – it will conflict with what your brain is trying to learn.

- When you go back to your revision, review what you have just learnt.
- Regularly review the facts you have learnt.

Get motivated

- Set yourself some goals and promise yourself a treat when the exams are over.
- Make the most of all the expertise and talent available to you at school and at home. If you do not understand something, ask your teacher to explain.
- Get organised. Find a quiet place to revise and make sure you have all the equipment you need.

Know what to expect in the exam

- Use past papers to familiarise yourself with the format of the exam.
- Make sure you understand the language examiners use.

Before the exam

- Have all your equipment and pens ready the night before.
- Make sure you are at your best by getting a good night's sleep before the exam.
- Have a good breakfast in the morning.
- Take some water into the exam if you are allowed.
- Think positively and keep calm.

During the examination

- Have a watch on your desk. Work out how much time you need to allocate to each question and try to stick to it.
- Make sure you read and understand the instructions and rules on the front of the exam paper.
- Allow some time at the start to read and consider the questions carefully before writing anything.
- Read all the questions at least twice. Do not rush into answering before you have a chance to think about it.
- If a question is particularly hard, move on to the next one. Go back to it if you have time at the end.
- Check your answers make sense if you have time at the end.

Tips for the science exam

- You should write your answers on the question paper.
- You may use a calculator.
- Remember that all questions should be attempted.
- Look at the number of marks allocated for each question in order to assess how many relevant points are required for a full answer. Very often, marks are awarded for giving your reasons for writing a particular answer.
- In numerical questions, workings out should be shown and the correct units used.
- Practical skills are important. Look back in your lab notes to remind yourself about why you carried out any practical work. What were you trying to find out? What did you actually do? What instrument did you use to take any measurements and what units did you use? How did you record your results: tables, bar charts or graphs? What were the results of your investigation and did you make any plans to change or improve what you did? These are all important and will be tested in the examination.
- A thorough understanding of your practical work will also help you to remember the key facts by putting them into context.
- Neat handwriting and careful presentation are important. When drawing diagrams, use a sharp pencil, and try to be as neat and precise as you can.

The organisation of living organisms

Know that living organisms are made of cells, the fundamental unit of living organisms

A living thing is called an **organism**. This term applies to any living thing, whether it is the smallest bacterium or the tallest tree.

All living organisms are made from **cells**. There are a great variety of cells. They are often constructed differently so they can carry out different tasks. However, they all have features that are common to all living organisms: **cytoplasm**, **cell surface membrane** (see Chapter 2).

Understand that living organisms are arranged as a hierarchy

cells → tissues → organs → systems

- **Cells**, generally of the same type and function, will combine together to make **tissues**: for example, muscle tissue or blood.
- **Tissues** of various kinds can combine to form an **organ**, which is a structure that performs a specific function: muscles create movement, skin protects and the **heart** pumps blood.
- Each **organ** has a specific job to do and in more complex animals such as mammals, organs will combine into a **system**.

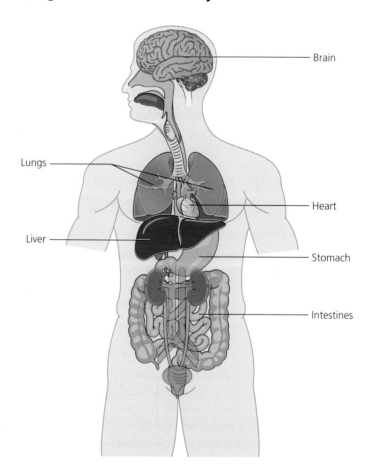

Brain

Lungs

Heart

Liver

Stomach

Intestines

■ Main organs in the human body

Systems **work together** to ensure the successful development and operation of the body.

Main systems in the human body

Name of system	What it does	Main organs in the system
Breathing system	Provides oxygen and removes carbon dioxide from the body	Windpipe, **lungs** and diaphragm
Digestive system	Breaks down food and absorbs useful chemicals into the blood	Stomach, liver and **intestines**
Reproductive system	Produces/receives sex cells for creating the next generation	Testes, ovaries, uterus
Nervous system (including sensory system)	Takes messages to/from all parts of the body via the **brain**. The brain co-ordinates the activities of different parts of the body	Brain, spinal cord, nerves, eyes, ears, nose, tongue
Muscle system	Allows movement	Muscles
Skeletal system	Supports and protects the body	Skeleton of bones
Urinary system	Filters waste products from the blood, producing urine and keeps the balance between water and salt in the body	Kidneys and bladder
Circulatory system	Moves blood around the body, pumps blood from the heart to the lungs to get oxygen	Heart and lungs

Main systems in plants

Plants also contain many cells that have specialised structures. The leaf, for example, is an organ that is made from various types of plant tissue. (See Chapter 2: Comparing plant and animal cells.)

Main organ of the system	Name of system	What it does
Flowers	Reproductive system	Flowers play an important part in reproduction
Stem	Transport system	Acts as a transport system, moving water and nutrients around. It holds the leaves towards the light and flowers up to insects and the wind
Roots	Transport system	Roots hold plants securely in the ground and absorb water and minerals from the soil
Leaves	Nutritional system	During photosynthesis, the nuclear store of energy in the Sun is transferred to a chemical store of energy in a plant by light. Energy is transferred to the green pigment, called **chlorophyll**, in the chloroplasts of cells in the leaves of the plant. (See Chapter 7).

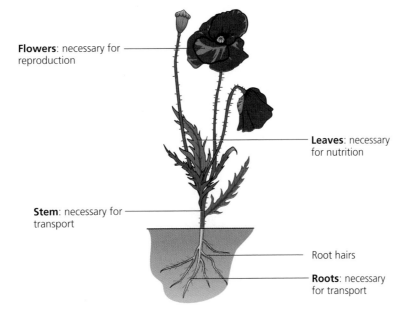

Flowers: necessary for reproduction

Leaves: necessary for nutrition

Stem: necessary for transport

Root hairs

Roots: necessary for transport

Recognise that each of these arrangements show the seven life processes

While there are huge numbers of different organisms, there will be processes that they carry out that are shared by each and every one of them. These **seven life processes** (sometimes called the characteristics of living things) work together to keep living organisms alive. Every cell in every tissue, in every organ and in every system needs to carry out these life processes to survive.

Movement

- Animals need to move from place to place to find food or escape from being someone else's food (see Chapter 11: The interdependence of organisms in an ecosystem). They will move to find a mate or to escape natural disasters (floods, fire, and so on).
- Plants also move: petals open and close, whole flower heads will turn to follow the path of the Sun, shoots respond to light and roots to gravity.

Respiration

This is the process that transfers energy from the chemical store of **glucose** to the energy stores of the cell. All living organisms need energy for movement, **growth** and repair.

Sensitivity

All living things react to changes their environment. For example, animals will react to temperature, sound and smell. Plants will move towards the light.

Growth

All organisms are made up from individual building blocks called cells. Growth is an increase in cell number and increase in cell size. Growth is how an organism becomes bigger and, sometimes, more complicated.

Reproduction

All organisms need to reproduce. As each individual organism will die sooner or later, it is vital that it is able to make more of its own kind, otherwise eventually species will become extinct. This is the way features (called characteristics) are carried forward from parents to the next generation (offspring). (See Chapter 6: Reproduction in humans and Chapter 9: Reproduction in flowering plants.)

Excretion

Respiration and other cell processes produce waste material, which needs to be removed. Animals remove waste through urine, faeces, sweat and exhaling carbon dioxide. Some plants expel waste accumulated in their leaves when the leaves fall in the autumn.

Nutrition

All organisms need to be able to obtain and absorb food that provides them with energy to live and grow.

- In animals, the process of digestion breaks down food. The energy released is transferred to the cells for growth and repair. If food is not consumed, it can result in the death of the animal.
- Plants produce their own food through the process of photosynthesis. If the key reactants (carbon dioxide, water and light) of photosynthesis are missing, the plant will die.

Recommended practical activities

Recall how you would observe, interpret and record how animal and plant cells are arranged into tissues and organs. Here are a few examples showing the hierarchical pattern in animals and plants.

Cells, tissues and organs in an animal

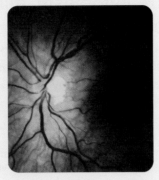

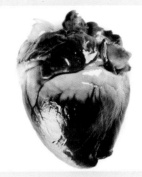

- A specialised **cell**: red blood cells

- Larger numbers of specialised cells join together to make up a **tissue**: blood vessels

- Various tissues together make up an **organ**: a heart

Cells, tissues and organs in a plant

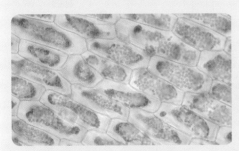

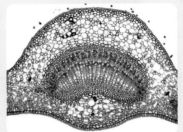

- This shows a specialised **cell** in a plant leaf which contains the green pigment, chlorophyll

- Larger numbers of specialised cells join together to make up leaf **tissue**: cross-section of a leaf viewed through a microscope

- Various tissues together make up a plant **organ** – the leaf

Exam-style questions: Exercise 1

1 a) Arrange the following in the correct hierarchy. (1)
    system   cell   organ   tissue
  b) Give an example for each one. (4)

2 Complete the following sentences, using the words provided.
  stomach   lungs   intestines   liver   brain   heart

  a) The _____ stores and digests food. (1)

  b) The _____ pumps blood around the body. (1)

  c) The _____ co-ordinates the activities of different parts of the body. (1)

  d) Digested food is absorbed in the _____ transferring nutrients into the bloodstream. (1)

  e) Oxygen enters the _____ when we breathe in. (1)

  f) The _____ removes poisons from the blood. (1)

3 Your skin is the largest and heaviest organ in your body. Explain why it relies on the life processes to thrive. (6)

Recognise that a typical animal or plant cell has: a nucleus, cytoplasm, mitochondria and cell surface membrane

Animal and plant cells have common features: **nucleus**, cytoplasm and a cell surface membrane. These features allow the cells to carry out the basic processes needed to stay alive.

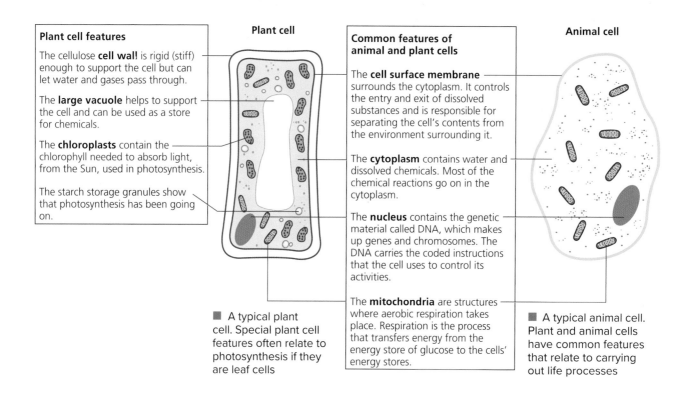

Plant cell features

The cellulose **cell wall** is rigid (stiff) enough to support the cell but can let water and gases pass through.

The **large vacuole** helps to support the cell and can be used as a store for chemicals.

The **chloroplasts** contain the chlorophyll needed to absorb light, from the Sun, used in photosynthesis.

The starch storage granules show that photosynthesis has been going on.

Plant cell

Common features of animal and plant cells

The **cell surface membrane** surrounds the cytoplasm. It controls the entry and exit of dissolved substances and is responsible for separating the cell's contents from the environment surrounding it.

The **cytoplasm** contains water and dissolved chemicals. Most of the chemical reactions go on in the cytoplasm.

The **nucleus** contains the genetic material called DNA, which makes up genes and chromosomes. The DNA carries the coded instructions that the cell uses to control its activities.

The **mitochondria** are structures where aerobic respiration takes place. Respiration is the process that transfers energy from the energy store of glucose to the cells' energy stores.

Animal cell

■ A typical plant cell. Special plant cell features often relate to photosynthesis if they are leaf cells

■ A typical animal cell. Plant and animal cells have common features that relate to carrying out life processes

Know the functions of each cell component

- **Nucleus:** contains information that the cell uses to make proteins grow and stay alive. The information is stored as **genes** in the **molecule DNA**. One of the main functions is the part it plays in reproduction. All the information needed for a cell to replicate itself is carried on chromosomes, which consist of genes, made from long strands of DNA.
- **Cytoplasm:** contains water and dissolved chemicals. Most of the chemical reactions go on in the cytoplasm.
- **Cell surface membrane:** encloses the cytoplasm, holding it together. It controls the passage of chemicals into and out of the cell.
- **Mitochondria:** structures where respiration takes place. **Respiration** is the process that transfers energy from the energy store of glucose to the cells' energy stores.

Recognise that plant cells contain cell walls, and often vacuoles and chloroplasts

In addition to the features animals and plant cells have in common, plant cells have a **cellulose cell wall**, **vacuoles** (and **chloroplasts** – only in cells such as leaf cells that photosynthesise).

Know the functions of each cell component

- **Cellulose cell wall:** is rigid, supporting the cell while allowing gases and water to pass through.
- **Vacuoles:** are fluid-filled spaces that support the cell and are a store for chemicals.
- **Chloroplasts:** contain the green pigment **chlorophyll**, which absorbs light during photosynthesis.

Recommended practical activities

1. Use a microscope to observe plant and animal cells.

Cells are too small to see unaided with the human eye, so a microscope can be used to help study them.

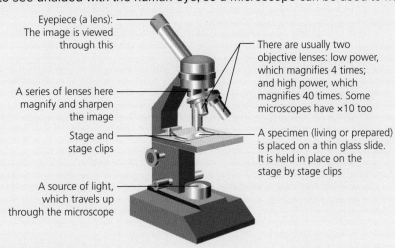

Eyepiece (a lens): The image is viewed through this

There are usually two objective lenses: low power, which magnifies 4 times; and high power, which magnifies 40 times. Some microscopes have ×10 too

A series of lenses here magnify and sharpen the image

Stage and stage clips

A specimen (living or prepared) is placed on a thin glass slide. It is held in place on the stage by stage clips

A source of light, which travels up through the microscope

- Place the slide on the stage.
- Make sure the lens is set to low power first.
- Move the lens as close to the specimen as possible – but not touching the slide.
- Look through the eyepiece, turn the focus knob towards you and wind the tube upwards until the specimen comes clearly into view.
- Use the fine focus knob (if fitted) to obtain a focused image.

2. Remember how to prepare a temporary stained microscope slide to observe animal and plant cells.

- When mounting specimens, it is usual to put the specimen (of cells) on a thin glass slide with a drop or two of mounting fluid – usually water.
- A cover slip is normally placed on top of the specimen to provide a flat surface to help with focusing the microscope.
- Stains are sometimes used to enable us to see certain features of a cell more clearly. Here are two you may have used:
  - Iodine is often used for looking at plant cells. A nucleus will appear orange-yellow; starch grains will appear **blue-black**.
  - Methylene blue is often used for looking at animal cells. The nucleus will appear blue.

3. Calculate the magnification of a light microscope

A light microscope uses the eyepiece and the objective lens to magnify the specimen.

The magnification of a lens is shown by a multiplication sign. So ×10 means the lens magnifies ten times.

To calculate the total magnification of a microscope, you use the following calculation:

magnification of a microscope = magnification of the eyepiece × magnification of the objective lens

The highest total magnification for a compound light microscope is ×1000

Exam-style questions: Exercise 2

1 Match the following cell features with the correct function. (6)

Feature		Function
A Vacuole		**1** Contains the green pigment, chlorophyll
B Chloroplast		**2** The structure that carries out aerobic respiration
C Chlorophyll		**3** Fluid-filled space found in plant cells
D Cytoplasm		**4** Common to both animal and plant cells
E Nucleus		**5** Contains genetic material called DNA, which makes up genes and chromosomes
F Mitochondria		**6** Absorbs light for photosynthesis

2 Draw and label a typical plant cell. (4)

3 Do all plant cells contain chloroplasts? Explain your answer. (2)

4 A microscope has an eyepiece with magnification ×10 and an object lens with magnification ×50. Calculate the total magnification of the microscope. (1)

5 A slide has been left on a microscope. You look at it and see that some of the cell visible on it is stained orange-yellow but there are little blue-black dots scattered around.

   a) What conclusion might you reach about whether it is an animal or a plant cell? (1)

   b) Explain your reasoning. (1)

   c) What other features might you see under the microscope? (4)

3 Food nutrients

Preliminary knowledge

All organisms need food to provide them with a store of energy and all of the chemical substances necessary to build the structure in cells. Humans, like all animals, need to eat food to supply the stores of energy for growth and repair of cells. During aerobic (using oxygen) respiration, energy in the chemical stores of energy in food is transferred to the cells of the body so growth and repair can take place.

- **Carbohydrates** are a chemical store of energy and include glucose and starch.
- **Proteins** are needed for growth and repair of cells.
- **Fats** are a chemical store of energy and provide insulation.

Preliminary practical activity

Remember how to test for starch using a solution of iodine. The iodine solution gives a dark blue-black colour when it is mixed with a food substance that contains starch.

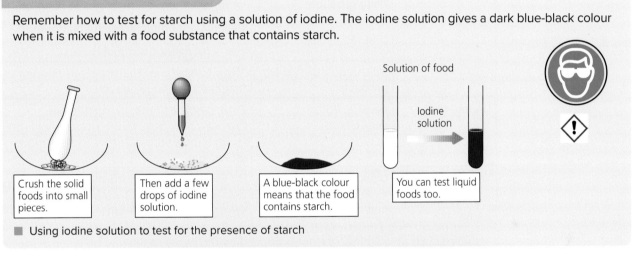

Solution of food

Iodine solution

| Crush the solid foods into small pieces. | Then add a few drops of iodine solution. | A blue-black colour means that the food contains starch. | You can test liquid foods too. |

■ Using iodine solution to test for the presence of starch

Test your preliminary knowledge: Exercise 3A

1   Match each word to the correct phrase. (4)

Food type	Function
A Fats	**1** Are carbohydrates that we should only eat in small quantities
B Proteins	**2** Is a food containing carbohydrate that is a good source of energy
C Wholemeal pasta	**3** Provide insulation
D Sugars	**4** Are needed to recover from an injury

2   Complete the following sentences:

a)   Food is essential to carry out the seven life processes: ____ , ____ , ____ , ____ , ____ , ____ and ____ . (7)

b)   ____ is used during ____ to transfer ____ stores of energy in food to the ____ of the body. (4)

c)   Too much ____ in your diet can cause tooth decay. (1)

d)   ____ solution gives a dark ____ colour when it is mixed with a food containing starch. (2)

Know the roles of carbohydrates, lipids (fats and oils), proteins, vitamins, minerals, dietary fibre and water in maintaining healthy bodies

A healthy diet includes a balanced selection of seven food types to provide us with all the nutrients we need, in the correct proportions.

Type of food	Where it is found	How it is used
Carbohydrates: Sugars (glucose) Starches	cakes, sweets, fruit bread, pasta, rice	Chemical energy store providing most of the energy we need. No more that 25% of the amount of this eaten should be sugars.
Fats	meat, butter, milk, cheese	Chemical energy store that can be used for respiration; it can provide insulation; too much can cause obesity and heart disease.
Proteins	fish, meat, milk, eggs	Growth and repair of cells.
Mineral salts	meat, vegetables, dairy products	Needed in small amounts to enable important chemical reactions to happen. Calcium – for healthy teeth and bones. Iron – for making red blood cells.
Vitamins*	fruit, vegetables, dairy products	Make certain chemical reactions happen that keep the body healthy.
Dietary **fibre**	cereals, fruit, vegetables	From plants and mainly indigestible. Provides bulk helping the passage of food move through the digestive system more effectively.
Water (70% of human body)	drinks, some foods especially salads and fruit	Dissolves and transports materials. You can lose about 1.5 litres of water a day as urine, sweat and in your breath.

* Vitamins such as vitamin C are not produced by the body, so have to be taken in by food. Lack of vitamin C causes bleeding gums.

Exam-style questions: Exercise 3B

1 Match each of the words in the first column with the phrase from the second column that will make a correct sentence. (6)

A Mineral – Iron	**1** can be caused by a lack of dietary fibre.
B Water	**2** is a good source of water.
C Benedict's solution	**3** will turn a solution containing protein purple.
D Constipation	**4** is good for making red blood cells.
E Cucumber	**5** is used to test for the presence of sugar.
F Biuret Reagent	**6** helps transport materials around the body.

2 a) Give an example of a food that would test positive for the presence of protein and fat. (1)

b) Describe how you would test this food for the presence of fat. (2)

3 Explain why it is so important to eat a lot of vegetables. (4)

(Recommended practical activities are on the following page.)

Recommended practical activities

Remember how to carry out tests for the presence of the following nutrients in foods:

1. Starch (see diagram at the start of the chapter)

Step 1   Crush solid foods into small pieces.

Step 2   Add a few drops of **iodine** solution.

Step 3   A colour change from brown to **blue-black** shows starch is present.

2. Sugar

Step 1   Crush and add water.

Step 2   Add an equal volume of **Benedict's solution** and heat the solution in a hot water bath.

Step 3   If sugar is present, a **red-orange** colour is seen.

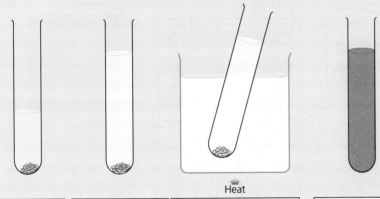

Heat

| Crush the food and dissolve it in water. | Add an equal volume of Benedict's solution. | Heat the mixture in a water bath (NOT over an open flame!). | A red-orange colour means that the food contains sugar. |

■ Using Benedict's solution to test for the presence of sugar

3. Fat (lipid) – the emulsion test

Step 1   Mix substance with **95% ethanol**.

Step 2   Add an equal volume of **distilled water**.

Step 3   If fat (lipid) is present, a milky-white emulsion forms.

4. Protein

Step 1   Crush and add water.

Step 2   Add an equal volume of **Biuret Reagent**.

Step 3   If protein is present, a **mauve-purple** colour is seen.

■ A positive test for the presence of fat using ethanol and distilled water. A cloudy emulsion forms at the top.

Biuret Reagent

■ The mauve-purple colour indicates a positive test for the presence of protein.

Understand how to interpret nutritional information on food labels

Nutrition labels are often found on packaged foods. The labels include information on the amount of energy stored in the food, together with information on the amount of fat, saturates (saturated fat), carbohydrates, sugars, proteins and salt in the food.

PER SERVING

ENERGY	FAT	SATURATES	SUGARS	SALT
1138kJ 272kcal	22.1g	5.5g	6.1g	0.2g
28%	32%	28%	7%	3%

OF YOUR GUIDELINE DAILY AMOUNT

Adobe Stock | #184630395

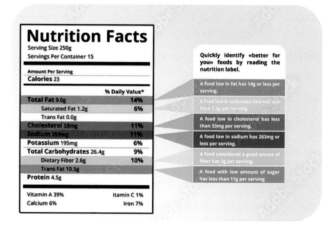

- **Some** food labels are simple, showing the amount (grams) of each food type and the percentage of your recommended daily allowance this represents.

- **Other** food labels contain more information: vitamins (A & C), minerals (calcium and iron).

Recognise foods that are a good source of each food group

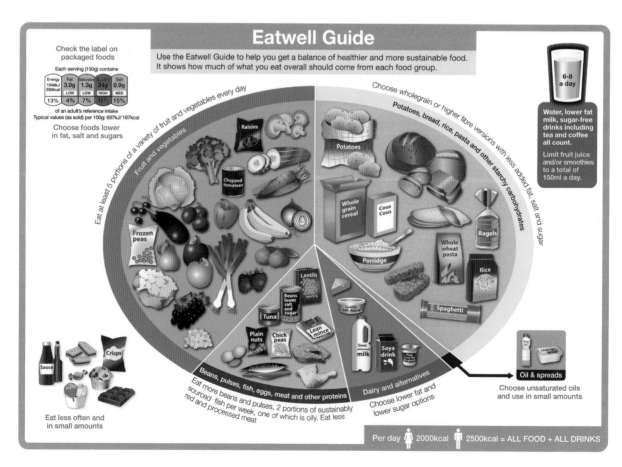

Recognise the consequences of imbalances in diet for healthy living

We need to have a mixture of foods to provide us with these nutrients in the right proportions. If we do not eat a balanced diet, it is likely to damage our health.

Here are some examples of the effects of imbalances:

Problems caused by a lack of something in our diet

Health problem	Caused by a lack of ...	What it is needed for	Where it is found
Scurvy (bleeding gums), poor growth in children, tissue repair is very slow	Vitamin C	Tissue repair, resistance to disease	fresh fruit, vegetables
Poor bone development, weak bones (osteoporosis)	Mineral – Calcium	Making bones and teeth, blood clotting	dairy products, flour, green vegetables
Shortage of red blood cells so less oxygen is transported around the body, making people tired (anaemia)	Mineral – Iron	Manufacture of red blood cells	meat, green vegetables
Constipation (faeces get stuck) and bowel cancer	Fibre	Enables food to move through the digestive system	cereals, wholegrain foods, vegetables

Problems caused by too much of something in our diet

Health problem	Caused by too much ...	What it is needed for	Where it is found
Obesity (large increase in body mass), heart disease, diabetes, joint damage	Fat	Chemical store of energy, insulation	meat, dairy products, fried food
Raised blood pressure, which can increase risk of heart disease and strokes	Mineral – Salt	To balance fluids in the body for healthy blood pressure	ready-made foods: pizzas, sausages, bacon, crisps
Tooth decay, obesity	Carbohydrate – Sugar (glucose)	Chemical store of energy	fruits, jams, soft drinks, sweets

Not eating enough food will mean there is not enough energy to carry out life processes. Starvation, reduction in body mass and eventually death will result.

Recommended practical activities

1. Food labels

Analyse food labels to investigate nutritional quality.

2. Food diary

You may have written a food diary to investigate dietary preferences. Revisit it and remind yourself of your findings.

Exam-style questions: Exercise 4

1 Choose which option is the best one to complete each of the following sentences.
   a) Calcium is essential in a healthy diet to _____ . (1)
   prevent constipation   develop strong teeth   be a source of energy

   b) Eating wholemeal foods will provide _____ , aiding digestion. (1)
   fibre   salt   water

   c) Eating too many _____ can lead to obesity.
   vegetables   processed foods   strawberries (1)

   d) The lack of vitamin C in a diet can cause _____ . (1)
   obesity   scurvy   raised blood pressure

2 The table below shows the relative amounts of different nutrients in food that might be eaten at breakfast.

Food	Carbohydrate, in g per 100 g	Fat, in g per 100 g	Protein, in g per 100 g
Orange juice	8	0	0
Bacon	0	10	12
Egg	0	5	6
Bread	24	Very small	4
Butter	0	8	Very small

   a) Name one food that is a good supply of energy. (1)

   b) Name two foods that are good for bodybuilding and growth. (2)

   c) Identify the main carbohydrate found in:
      i) orange juice
      ii) bread (2)

   d) Name two important components of a healthy diet not included in the table. (2)

3 The following table shows the average daily amounts of protein that (2)
   human males of different ages require if they are to remain fit and healthy.

Age, in years	11	14	18	25	45	65
Protein requirement, in g per day	75	85	100	65	65	65

   Explain the differences in protein requirements for human males at ages 18 and 45.

4

	Snack X	Snack Y
Fat	22%	34%
Fibre	18%	35%
Carbohydrate	60%	31%

   a) i) Which snack contains the greatest amount of fibre? (1)
      ii) Suggest why fibre is an important part of your diet. (1)

   b) i) Which snack would you eat before running a race? (1)
      ii) Give a reason for your answer. (1)

5 Vitamins and minerals are important parts of our diets.
   Give one example of each, suggesting what they do. (4)

Ask yourself

The southern part of the island nation of Madagascar, off the east coast of Africa, is experiencing its worst drought in four decades, with the World Food Program (WFP) warning recently that 1.14 million people are food-insecure and 400 000 people are headed for famine. Hunger is already driving people to eat raw cactus, wild leaves and locusts, a food source of last resort. … [A] WFP spokesperson [says], "The land is covered by sand; there is no water and little chance of rain."

Time, 20 July 2021

Use what you know to consider how this situation will affect the health of the Madagascan people. What health issues might they face?

■ Baobab trees in Madagascar

The structure of the lungs

- There are two lungs inside the ribcage.
- Breathing is the movement of air in and out of the lungs.
- The lungs take oxygen from the air and pass it into our blood. At the same time, they remove waste carbon dioxide from our blood and pass it into the air. This is known as gaseous exchange.
- The lung surface is greatly folded, creating a large surface area for gaseous exchange.
- The oxygen that is taken into the lungs by breathing is transported in the bloodstream to the tissues by the circulatory system.

Damage to the lungs

- Smoking is one of the causes of lung cancer and heart disease.
- Tar in tobacco smoke covers the surface of lungs, reducing the surface area across which gases can be exchanged. This can lead to severe breathing difficulties.
- Smoking can also cause cancer and heart disease.

Test your preliminary knowledge: Exercise 5A

1 Complete the sentences using the words below. You may use the given words more than once.

bloodstream   oxygen   carbon dioxide   lungs   nose

a) _____ enters the body through the _____ and travels down into the _____ .   **(3)**

b) The _____ transports _____ to the tissues of the body.   **(2)**

c) Blood is pumped to the _____ where _____ then leaves the body.   **(2)**

2 Explain how the lungs are designed to increase the rate of gas exchange.   **(1)**

3 Explain how the tar in tobacco smoke causes breathing difficulties.   **(1)**

Know how the movements of the diaphragm and ribcage lead to breathing in humans

- The main organs for breathing in humans are the **lungs**, which are located in the **chest** cavity and protected by the **ribcage**.
- The chest is lined with a membrane and the floor of the chest is separated from the abdomen by a flexible membrane called the **diaphragm**.
- There are two sets of **intercostal muscles** located between the ribs and two sets of diaphragm muscles.
- As the intercostal muscles contract, the ribcage moves upwards and outwards. As the diaphragm muscles contract, the diaphragm moves downwards.
- The volume of the air in the chest cavity increases and the pressure of the air decreases. The internal air pressure falls below the air pressure outside and air is drawn into the lungs through the nose and **trachea**.
- As these intercostal muscles relax, the ribcage moves downwards and inwards. As this diaphragm muscle relaxes, the diaphragm moves upwards.
- The volume of air in the chest cavity falls and the pressure increases. The internal air pressure rises above the pressure outside and air leaves the lungs through the nose and trachea.

Understand how changes in pressure lead to breathing in and out in humans

- The movements of the ribcage and the diaphragm cause the pressure of the air inside the lungs to change relative to the air pressure outside.
- When the air pressure inside the lungs is **lower** than the air pressure outside, air rushes into the lungs. This is called **inhalation**.
- When the air pressure in the lungs is **greater** than the air pressure outside, air is blown out of the lungs. This is called **exhalation**.

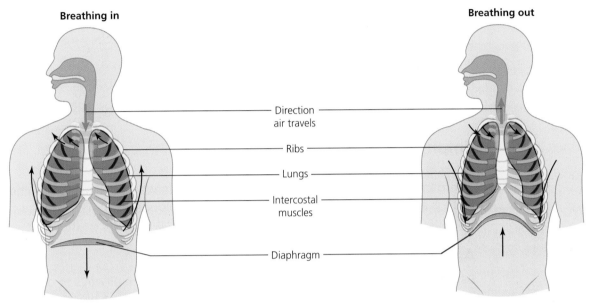

Breathing in

Breathing out

Direction air travels

Ribs

Lungs

Intercostal muscles

Diaphragm

The diaphragm and intercostal muscles contract. This increases the volume of the chest cavity so air is pushed into the lungs.

The diaphragm and intercostal muscles relax. This reduces the volume of the chest cavity so air is pushed out of the lungs.

Recognise that medical conditions such as emphysema and asthma reduce the movement of oxygen into the lungs

It is worth reminding yourself of the composition of air:

	Approximate content of inhaled air	Approximate content of exhaled air
Nitrogen	78%	80%
Oxygen	20%	16%
Carbon dioxide	0.04%	4%

Recall the route air takes to enter the lungs:

nose → throat → trachea → bronchus → bronchiole → air sacs in the lungs

Oxygen comes from the inhaled air. The air sacs are surrounded by blood vessels and this is where the oxygen enters the blood stream via a process known as gas exchange.

Problems with breathing arise if the route taken by air to reach the air sacs in the lungs is impeded in any way.

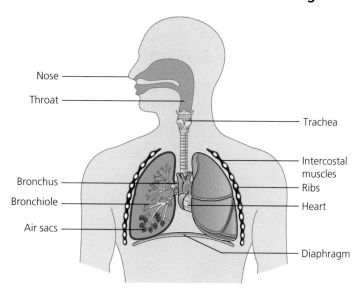

Emphysema

This is a condition where the air sacs in the lungs are damaged. This makes it difficult to breathe and there is a struggle to get enough oxygen into the bloodstream. The air sacs can be damaged by smoking or by being exposed to damaging particles in the air.

Asthma

Asthma is a medical condition that causes muscles in the airways to contract, making it difficult to breathe and get enough oxygen into the bloodstream. Air cannot move easily in and out of the lungs. Asthma is caused by a variety of factors, including anxiety, pollen, household dust and air pollution. Use of an inhaler that blows a spray of chemicals to relax the bronchial muscles is one method of providing relief.

Recommended practical activities

1. **Demonstrate, using the bell jar and parallelogram models, the role of the diaphragm and ribs in breathing.**

   a) Bell jar model to illustrate the role of the diaphragm in breathing

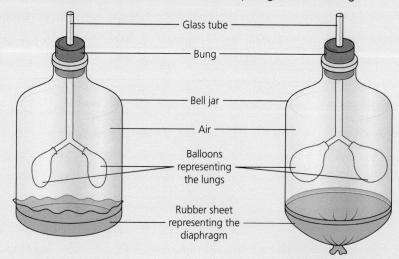

- The rubber sheet represents the diaphragm. Before it is pulled, the balloons are deflated.

- When the rubber sheet is pulled, simulating the contraction of the diaphragm, air is pulled in via the glass tube inflating the balloons.

b) Parallelogram model to demonstrate the role of the ribs in breathing.

- The model consists of wooden pieces, arranged as shown below. The horizontal pieces, representing the rib bones, should be the same size.
  The two pieces representing the backbone and the sternum are of different sizes. The pieces are loosely connected by screws and nuts, so that the pieces can all move upwards and downwards.
- Attaching the elastic band to P and Q causes the ribs to move upwards.
  At this point, the elastic band cannot move back to its original position without being stretched.
- Attaching an identical band at R and S will allow this to happen and the ribs to move downwards.
- The elastic bands represent the intercostal muscles that are found on either side of each rib bone.
- The movement of the ribs up and down requires two sets of muscles. This is because muscles can only get shorter. They cannot return to their original lengths on their own.
- The muscles around the ribs act together. They are called **antagonistic pairs** because they act in opposition to each other.

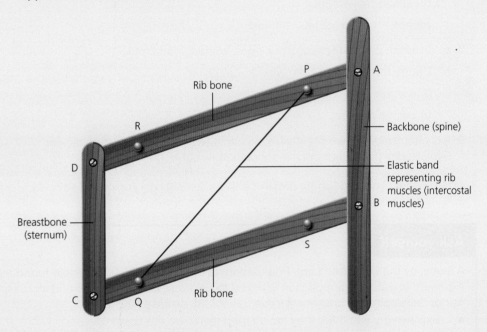

2. Measure vital capacity in humans, using peak flow meters.

Vital capacity is the maximum amount breathed in with the deepest breath.

You can measure how much air you breathe out by using a **peak flow meter**.

Step 1   Check the indicator on the peak flow meter shows zero.

Step 2   Take a deep breath. Put the peak flow meter in your mouth and seal it with your lips so no air escapes.

Step 3   Blow out hard and fast. You will see the indicator move in response.

Step 4   The number the indicator is pointing to is a measure of your vital capacity.

■ Blowing into a peak flow meter to measure vital capacity

You should carry out this test three times and record the best result.

Exam-style questions: Exercise 5B

1 Choose which option best completes each of the following sentences.

a) Breathing is defined as the movement of _____ in and out of the lungs. (1)

intercostal muscles **air** **just oxygen** **just carbon dioxide**

b) When the air pressure in the lungs is greater than the pressure of the air outside the body, air moves _____ the lungs. (1)

out of **into**

c) Gas exchange in humans takes place in the _____ . (1)

trachea **bronchiole** **nose** **air sacs**

d) _____ is a disease where the air sacs in the lungs are damaged. (1)

Scurvy **Emphysema** **Diabetes**

e) The _____ muscles connecting the ribs are responsible for the breathing movement. (1)

triceps **intercostal** **heart**

2 a) In which part of the lungs does gas exchange take place? (1)

b) During this process, state what happens to:

i) oxygen (1)

ii) carbon dioxide (1)

3 It is often said that we breathe in oxygen and breathe out carbon dioxide. Explain why this statement is not completely correct. (2)

4 How would you test for the presence of carbon dioxide in exhaled air? (1)

Ask yourself

A study, by the European Lung Foundation, of more than 300 000 people found that exposure to outdoor air pollution is linked to decreased lung function and increased risk of developing chronic obstructive pulmonary disease (COPD).

● Explain why you think they have found this to be the case.

● Explain what can happen to the lungs of people exposed to air pollution.

● What health issues might they experience?

6 Reproduction in humans

Reproduction is one of the seven life processes. It occurs in all living things, plants and animals. **Fertilisation** is the fusing of male and female sex cells in sexual reproduction.

Reproduction in plants and animals

Sexual and asexual reproduction in plants

- **Sexual** reproduction in a plant happens when the male and female sex cells, contained in the flower, join together to make seeds. (See Chapter 9: Reproduction in flowering plants.)
- It is possible for plants to reproduce **asexually**; no fertilisation takes place in this process. Examples of this are reproduction through bulbs and tubers. Energy stored in the bulbs and tubers provides the energy for cell division and growth.

Sexual reproduction in animals

Sexual reproduction in animals occurs when the male and female sex cells fuse to form an embryo.

Main stages of the human life cycle

- Fertilised egg: made by the joining of specialised cells, called gametes, from the mother and father.
- Embryo: development of the new human inside the mother.
- Baby: cannot feed or walk without help.
- Child: can feed, walk and talk but is still reliant on parents.
- Adolescent: sex organs mature but growth continues.
- Adult: growth complete; reproduction is possible.
- Old age: no growth; some life processes work less well.
- Death: life processes end.

Adolescence

Time of physical change

- Occurs between the ages of 10 and 20.
- Body hair starts growing around the genitals (pubic hair) and on other parts of the body.
- In males, hair grows on the chest and face, and voice becomes deeper.
- In females, the menstrual cycle begins.
- Reproductive organs develop: in males, the penis becomes larger; in females, breasts develop and hips become wider.

Time of emotional change

- Young people become more independent and more responsible for their thoughts and actions.
- They develop strong feelings of sexual attraction.

Test your preliminary knowledge: Exercise 6A

1 Match each stage in the human life cycle with the correct description. **(4)**

Stage	Description
A Fertilisation	**1** A period of physical and emotional change occurring when humans are between 10 and 20 years old.
B Embryo	**2** Needs help to feed and survive.
C Baby	**3** The fusion of the egg and the sperm.
D Adolescence	**4** Ball of cells which becomes embedded in the uterus.

2 Generally, animal size and mass contribute to the gestation period (the time between fertilisation and birth). Larger animals tend to have longer gestation periods. Use the words below to complete the table. **(5)**

elephant   21 days   cat   280 days   horse

Animal	Average gestation period (days)
Rat	**a)**
b)	63 days
Human	**c)**
d)	336 days
e)	624 days

Know the names and functions of the organs in the male and female reproductive systems

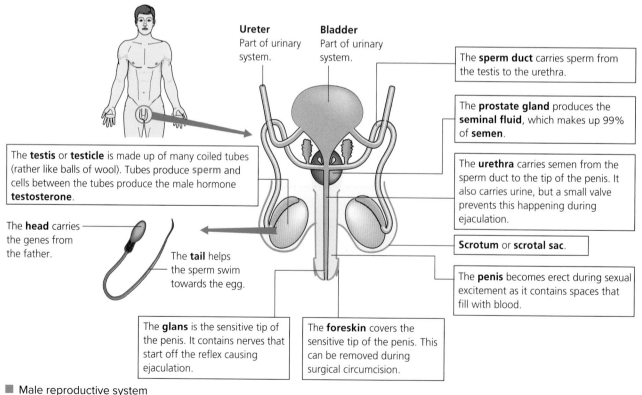

Ureter
Part of urinary system.

Bladder
Part of urinary system.

The **sperm duct** carries sperm from the testis to the urethra.

The **prostate gland** produces the **seminal fluid**, which makes up 99% of **semen**.

The **urethra** carries semen from the sperm duct to the tip of the penis. It also carries urine, but a small valve prevents this happening during ejaculation.

Scrotum or **scrotal sac**.

The **penis** becomes erect during sexual excitement as it contains spaces that fill with blood.

The **testis** or **testicle** is made up of many coiled tubes (rather like balls of wool). Tubes produce **sperm** and cells between the tubes produce the male hormone **testosterone**.

The **head** carries the genes from the father.

The **tail** helps the sperm swim towards the egg.

The **glans** is the sensitive tip of the penis. It contains nerves that start off the reflex causing ejaculation.

The **foreskin** covers the sensitive tip of the penis. This can be removed during surgical circumcision.

■ Male reproductive system

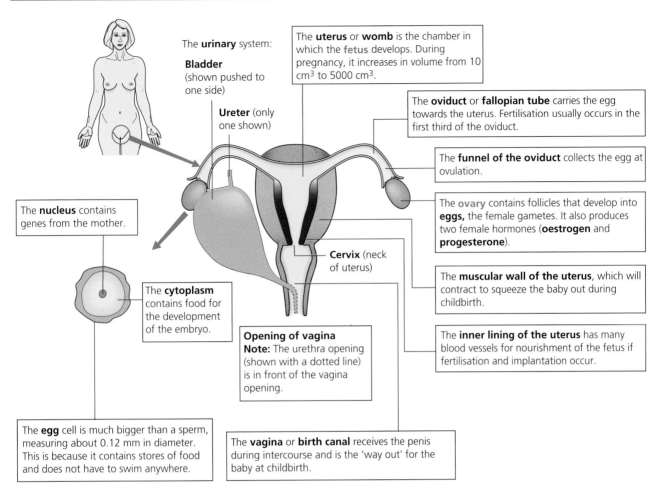

The **urinary** system:

Bladder (shown pushed to one side)

Ureter (only one shown)

The **uterus** or **womb** is the chamber in which the **fetus** develops. During pregnancy, it increases in volume from 10 cm³ to 5000 cm³.

The **oviduct** or **fallopian tube** carries the egg towards the uterus. Fertilisation usually occurs in the first third of the oviduct.

The **funnel of the oviduct** collects the egg at ovulation.

The **ovary** contains follicles that develop into **eggs,** the female gametes. It also produces two female hormones (**oestrogen** and **progesterone**).

The **nucleus** contains genes from the mother.

Cervix (neck of uterus)

The **cytoplasm** contains food for the development of the embryo.

The **muscular wall of the uterus**, which will contract to squeeze the baby out during childbirth.

The **inner lining of the uterus** has many blood vessels for nourishment of the fetus if fertilisation and implantation occur.

Opening of vagina
Note: The urethra opening (shown with a dotted line) is in front of the vagina opening.

The **egg** cell is much bigger than a sperm, measuring about 0.12 mm in diameter. This is because it contains stores of food and does not have to swim anywhere.

The **vagina** or **birth canal** receives the penis during intercourse and is the 'way out' for the baby at childbirth.

■ The female reproductive system

Understand how these parts work together to lead to fertilisation

- During sexual intercourse, special spongy tissue (erectile tissue) within the **penis** becomes full of blood, which enables the penis to become stiff enough to enter the female's **vagina**.
- Movement of the penis within the vagina causes a nervous reflex that is a signal for 300 million **sperms (male gametes)** contained in about 4–5 cm³ of fluid to be ejaculated out of the penis into the vagina.
- One egg (**ovum** – the **female gamete**) is released by the female every 28 days and is carried by cilia into the **oviduct**.
- Only a small proportion of sperms will complete the journey to the oviduct and it is here that only one sperm will fuse with the egg (**fertilise** it) to form a **zygote**.
- The zygote contains genes from both the mother and the father.
- The zygote begins cell division until a ball of about 128 cells (now called an **embryo**) arrives in the uterus and settles deep in the newly formed thick lining.
- This settling process is called **implantation**.
- Once the embryo is implanted, the mother is pregnant.

Recognise the importance of the menstrual cycle

The menstrual cycle prepares the body for pregnancy every month. An egg is released and the uterus is prepared for pregnancy if the egg is fertilised. If not, the lining breaks down in preparation for the next month's cycle.

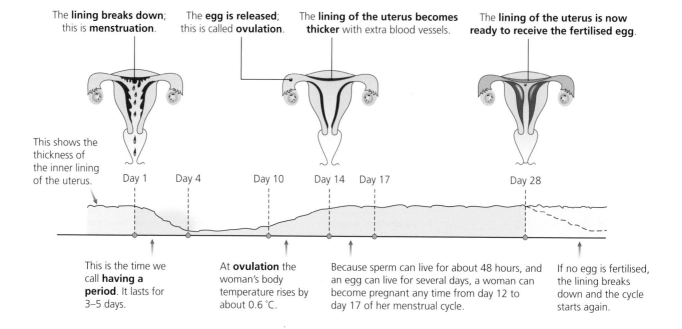

The **lining breaks down**; this is **menstruation**.

The **egg is released**; this is called **ovulation**.

The **lining of the uterus becomes thicker** with extra blood vessels.

The **lining of the uterus is now ready to receive the fertilised egg**.

This shows the thickness of the inner lining of the uterus.

Day 1   Day 4   Day 10   Day 14   Day 17   Day 28

This is the time we call **having a period**. It lasts for 3–5 days.

At **ovulation** the woman's body temperature rises by about 0.6 °C.

Because sperm can live for about 48 hours, and an egg can live for several days, a woman can become pregnant any time from day 12 to day 17 of her menstrual cycle.

If no egg is fertilised, the lining breaks down and the cycle starts again.

Recognise that the lifestyle of parents can affect the healthy development of a fetus

To recognise how the development of a fetus can be affected, it is necessary to understand what happens after implantation.

Development of the embryo

- The implanted embryo develops into the fetus. There is an increase in the number of cells and the cells develop into the different kinds of cell found in the body.
- The fetus is contained within an **amniotic sac** filled with **amniotic fluid**. It is attached to the placenta by the umbilical cord. The amniotic fluid keeps the fetus moist and protects it from physical knocks.

The placenta

- The placenta is a plate-shaped organ that grows deep into the uterus wall and increases in size as the fetus develops.
- It enables food and oxygen to be passed from the mother to the fetus. Carbon dioxide and waste materials pass from the fetus to the mother.
- Although the mother's blood vessels flow close to the blood vessels of the fetus, the two are kept separate. The blood of the mother and the blood of the fetus do not mix. This is because:
  - the blood group of the fetus may be different from that of the mother
  - bloods of different groups must not be mixed together.
- The mother's blood pressure will be much higher than that of the fetus.

The umbilical cord

- The cord contains blood vessels that:
  - take oxygen and food to the fetus
  - take carbon dioxide and nitrogenous waste away from the fetus.
- It is clamped at birth to prevent bleeding and then cut. The 'belly button' is the remains of the umbilical cord.

Pregnant mothers should be careful about their lifestyle and diet. This is because harmful substances can cross from the mother through the umbilical cord to the developing baby.

- Smoking can affect the fetus in two ways: the nicotine is addictive and poisonous gases may reduce the baby's birth weight.
- Alcohol can cause brain damage to the baby.
- Viruses such as HIV and sexually transmitted diseases can infect the baby; other diseases could result in harmful materials being passed to the fetus through the umbilical cord.
- Some foods can cause an infection that may lead to miscarriage. For example, unpasteurised products may contain listeria; toxoplasmosis may be picked up from raw meat.
- Excess physical activity may also harm the fetus.

Recommended practical activity

Study the development of a fertilised egg into a fetus.

 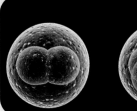  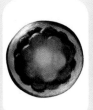 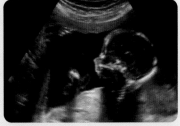

- A sperm fusing with an egg at the moment of fertilisation
- A zygote (a fertilised cell) starts to divide, first into two cells, then four and so on
- A ball of cells ready for implantation
- A developing human fetus seen by ultrasound

Exam-style questions: Exercise 6B

1 a) Name the parts of the male reproductive system labelled in the diagram below. (3)

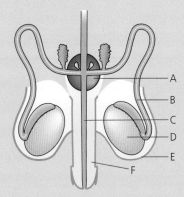

A
B
C
D
E
F

b) Give a short description of the function of each part. (6)

2 a) Name the parts of the female reproductive system labelled in the diagram below. (3)

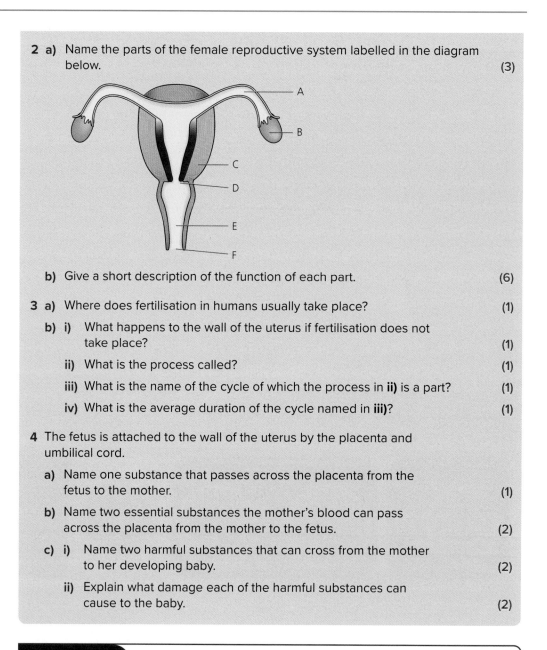

b) Give a short description of the function of each part. (6)

3 a) Where does fertilisation in humans usually take place? (1)

b) i) What happens to the wall of the uterus if fertilisation does not take place? (1)

ii) What is the process called? (1)

iii) What is the name of the cycle of which the process in **ii)** is a part? (1)

iv) What is the average duration of the cycle named in **iii)**? (1)

4 The fetus is attached to the wall of the uterus by the placenta and umbilical cord.

a) Name one substance that passes across the placenta from the fetus to the mother. (1)

b) Name two essential substances the mother's blood can pass across the placenta from the mother to the fetus. (2)

c) i) Name two harmful substances that can cross from the mother to her developing baby. (2)

ii) Explain what damage each of the harmful substances can cause to the baby. (2)

Ask yourself

Over the past few years, increases in cases of microcephaly (a condition where a baby is born with a head that is much smaller than those of other babies of the same age and sex) have been reported. From the available evidence, it has been suggested that infection with the Zika virus, a virus transmitted primarily by mosquitoes, during pregnancy is a possible cause of brain abnormalities, including microcephaly.

Why might a baby be affected in this way if their mother is bitten by a mosquito?

Identify and describe different parts of flowering plants

- **Roots: Hold** the plant firmly in the soil and **absorb water** and **minerals** from the soil. Minerals (for example, nitrates) are nutrients needed for healthy growth.
- **Stem: Supports** branches, leaves, flowers and fruits. It **transports** water and minerals from the roots to all parts of the plant. It moves food produced in the leaves to the growing and storage places of the plant.
- **Leaves:** Each leaf is a miniature factory making the food that the plant needs in order for it to grow. Cells inside the leaf contain a green pigment called **chlorophyll**. Chlorophyll absorbs the light needed for the plant to make food by the process called photosynthesis.
- **Flowers:** The flower contains the reproductive organs of a flowering plant. Petals surround the parts of the plants that make the gametes (female gametes, called ovules, and male gametes contained in pollen grains).

The effects of light, air, water and temperature on plant growth

To grow well, plants need the right conditions:

- Enough **light** to make food. The more light there is, the greater the rate of photosynthesis.
- **Water** is needed for photosynthesis and to carry food around the plant.
- **Carbon dioxide** from the air is essential for photosynthesis.
- The right **temperature**. Heat is needed so that chemical reactions can take place.

Remember:

- the effect of changes in light, temperature and water on plant growth
- that air supplies a plant with carbon dioxide for making food
- that plants also need oxygen.

Four similar plants are selected. Each one is treated in the same way as the control plant but one condition is changed in each case.

Condition changed	Situation	Result	Reason
No light	Dark cupboard	Plant grows tall and weak and the leaves turn yellow	Light is needed to make food
No warmth	In a fridge	Plant does not grow and leaves may become damaged	The plant needs warmth so that chemical reactions, needed for growth, can take place
No water	Not watered	Leaves wilt	Water is needed for photosynthesis and to carry food around the plant

Test your preliminary knowledge: Exercise 7A

1 Match each of the organs listed below to its numbered label. (4)
   flower   stem   root   leaf

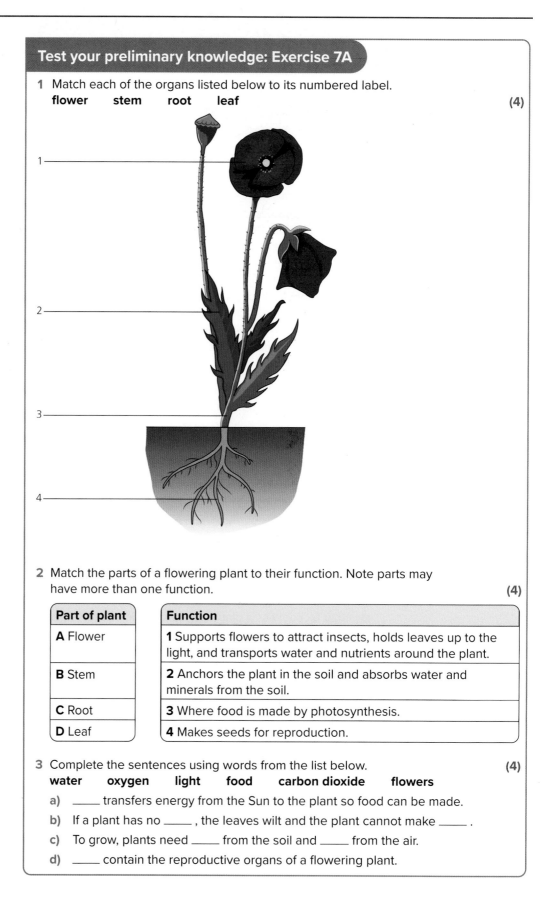

2 Match the parts of a flowering plant to their function. Note parts may have more than one function. (4)

Part of plant	Function
A Flower	**1** Supports flowers to attract insects, holds leaves up to the light, and transports water and nutrients around the plant.
B Stem	**2** Anchors the plant in the soil and absorbs water and minerals from the soil.
C Root	**3** Where food is made by photosynthesis.
D Leaf	**4** Makes seeds for reproduction.

3 Complete the sentences using words from the list below. (4)
   water   oxygen   light   food   carbon dioxide   flowers

   a) _____ transfers energy from the Sun to the plant so food can be made.

   b) If a plant has no _____ , the leaves wilt and the plant cannot make _____ .

   c) To grow, plants need _____ from the soil and _____ from the air.

   d) _____ contain the reproductive organs of a flowering plant.

Know the word equation for photosynthesis

Plants make their own food by the process of photosynthesis. They use **carbon dioxide** from the air and **water** from the soil. A plant uses energy transferred by **light** from the Sun to the **chlorophyll** in the leaves to produce sugars that are useful to all living organisms. The store of energy in the plant enables it to grow larger, resulting in an increase in biomass.

The process of photosynthesis can be summarised by the following word equation:

$$\text{carbon dioxide + water} \xrightarrow[\text{chlorophyll}]{\text{light}} \text{glucose + oxygen}$$

(reactants)   (products)

Carbon dioxide and water are the reactants and glucose (a carbohydrate) and oxygen are the products of photosynthesis.

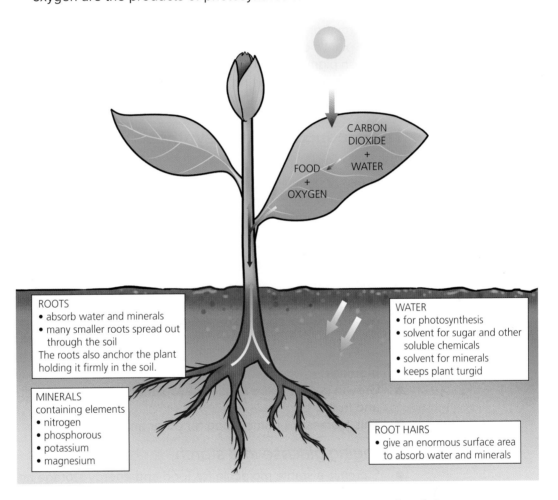

ROOTS
- absorb water and minerals
- many smaller roots spread out through the soil

The roots also anchor the plant holding it firmly in the soil.

MINERALS
containing elements
- nitrogen
- phosphorous
- potassium
- magnesium

CARBON DIOXIDE + WATER

FOOD + OXYGEN

WATER
- for photosynthesis
- solvent for sugar and other soluble chemicals
- solvent for minerals
- keeps plant turgid

ROOT HAIRS
- give an enormous surface area to absorb water and minerals

Understand the importance of carbon dioxide, water and light in the production of new biomass

The reactants of photosynthesis

Photosynthesis uses carbon dioxide, water and light to produce carbohydrates (sugars, such as glucose). Glucose is a chemical store of energy for growth and an increase in biomass. The carbohydrates made in photosynthesis provide more than just a store of energy. The atoms themselves are used to make glucose, starch and cellulose, all of which are essential parts of plant cells; the carbon and hydrogen atoms are also used to make proteins and fats.

Carbon dioxide

- Carbon dioxide supplies carbon and oxygen – some of the ingredients needed for making carbohydrates.
- It is absorbed from the air through the leaves. This gas is a product of respiration.

Water

- Water supplies hydrogen – an ingredient needed for making carbohydrates.
- It is absorbed from the surrounding soil through the plant's root system.
- It is a solvent for sugar and other soluble chemicals so enables them to move around the plant easily.
- It is a solvent for minerals, enabling them to be transported to the place where they are made into the proteins and chemicals that are essential to the growth and development of the plant.

Light

- Energy from the nuclear store of energy in the Sun is transferred by **light** to the plant's chemical store of energy in the **chloroplasts**, which contain the green pigment chlorophyll.
- Chloroplasts containing chlorophyll are found in the cytoplasm of nearly every leaf cell.

The products of photosynthesis

Oxygen

- Some of this will be used by the plant itself, for **aerobic respiration** (see Chapter 8: Respiration).
- Oxygen not used by the plant will be released to the air through the underside of the leaf.

Glucose for making living material – increasing biomass

- Photosynthesis provides food, in the form of glucose, for the plant. Energy stored in the food is either used immediately for growth or it is transferred to other chemical stores of energy such as starch.
- For **growth**, plants need a constant supply of proteins and fats, which can be made from sugars such as glucose. Plant growth causes an increase in **biomass**.
- Growth comes in the form of new plant cells and the production of seeds and fruit.
- This biomass becomes food for other animals.

Changing glucose into starch

- Starch is a stable molecule for food and energy storage and for making plant structures.
- In good light (in the daytime), glucose is made at a faster rate than the rate at which it can be transported for growth, so this excess glucose is changed into starch.
- In darkness (at night), starch in the leaves is changed back into glucose and transported to other parts of the plant.
- Food, in the form of starch, is stored in preparation for the growth of the next generation:
  - in seeds (peas and beans)
  - in storage organs, such as bulbs (onions) and tubers (potatoes).

Glucose for respiration

- Photosynthesis provides the glucose which every living cell needs for respiration.
- The leaves make more glucose than they need, so glucose is transported to other parts of the plant for respiration. (See Chapter 8: Respiration.)

Recognise the importance of photosynthetic organisms as producers in food chains

- The biomass of plants provides animals with a food supply.
- Plants are at the very start of all **food chains**/webs – this is why they are called **producers**. (See Chapter 11: The interdependence of organisms in an ecosystem.)
- Animals cannot carry out photosynthesis, so they have to obtain their energy by eating plants, or by eating other animals that have eaten plants. Energy is transferred from the chemical stores of energy in plants to the chemical stores of energy in animals.
- Without plants, food chains would collapse and animals would cease to exist.
- The remains of animals that have eaten biomass can form fuels such as coal, oil and gas. When these fuels are burned, the energy transferred can be useful to humans (by, for example, generating electricity, light and heat).

Recognise the importance of photosynthesis in maintaining the concentrations of oxygen and carbon dioxide in the atmosphere

Carbon dioxide is needed for photosynthesis and oxygen is needed for respiration. A balance between these two gases is necessary for life on Earth to continue.

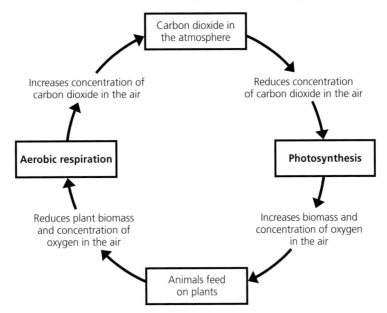

- Before the arrival of green plants on Earth, there was no oxygen at all in the atmosphere.
- All living cells in plants and animals need oxygen for respiration. It enables energy to be released from food.
- **Photosynthesis** produces more oxygen than the plant needs for **respiration**, so excess oxygen is released into the atmosphere through little holes on the underside of leaves.

- As a result of the evolution and development of green plants, the amount of oxygen in the atmosphere has built up to its present level of about 20% of all the gases in the air. So every leaf and blade of grass helps to maintain the amount of oxygen in the air.
- Oxygen needs to be replaced because respiration by all organisms (including plants) removes vast quantities of oxygen from the air every minute. If this oxygen were not replaced by photosynthesis, then all the oxygen would be removed from the air in a few thousand years.

Human actions can affect the balance of gases in the air

- Cutting down forests *reduces* the amount of photosynthesis that can occur, so the amount of carbon dioxide in the air increases because less is being used up.
- Burning fossil fuels *increases* the amount of carbon dioxide in the air.

Recommended practical activities

1. Investigate photosynthesis in variegated plant leaves using a starch test.

A variegated leaf (a leaf where part is green and part is not – often pale yellow) shows that chlorophyll is necessary for photosynthesis. Only the green parts containing chlorophyll can produce starch.

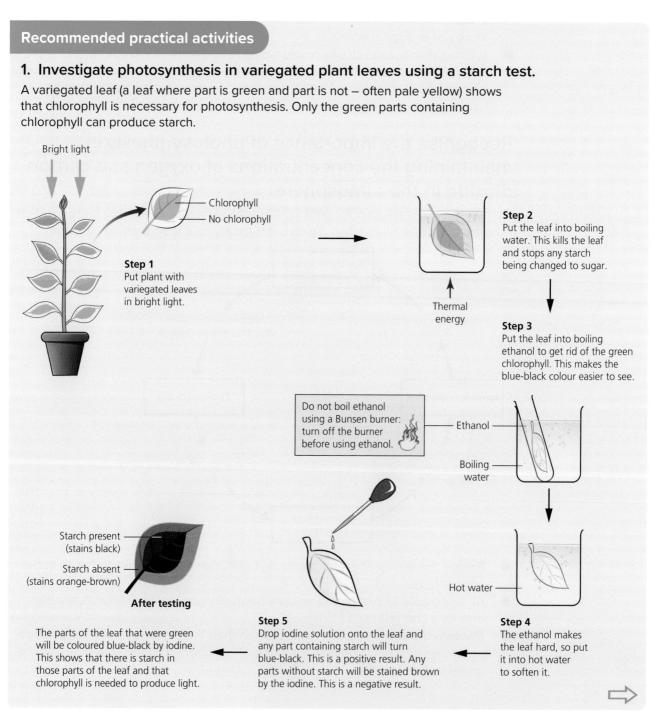

Bright light

Chlorophyll
No chlorophyll

Step 1
Put plant with variegated leaves in bright light.

Step 2
Put the leaf into boiling water. This kills the leaf and stops any starch being changed to sugar.

Thermal energy

Step 3
Put the leaf into boiling ethanol to get rid of the green chlorophyll. This makes the blue-black colour easier to see.

Do not boil ethanol using a Bunsen burner: turn off the burner before using ethanol.

Ethanol

Boiling water

Starch present (stains black)
Starch absent (stains orange-brown)

Hot water

After testing

The parts of the leaf that were green will be coloured blue-black by iodine. This shows that there is starch in those parts of the leaf and that chlorophyll is needed to produce light.

Step 5
Drop iodine solution onto the leaf and any part containing starch will turn blue-black. This is a positive result. Any parts without starch will be stained brown by the iodine. This is a negative result.

Step 4
The ethanol makes the leaf hard, so put it into hot water to soften it.

2. Investigate the effect of light intensity on the rate of photosynthesis in a Cabomba plant (or other aquatic plant).

This investigation can be used to test how fast photosynthesis is occurring. During photosynthesis, glucose is produced from carbon dioxide and water, and oxygen is produced as a waste product. Oxygen bubbles can be seen in the water as they are released from the plant. Counting the bubbles that appear in a measured amount of time (30 seconds or 1 minute) will tell us the rate of photosynthesis. These bubbles can be tested to prove the gas is oxygen.

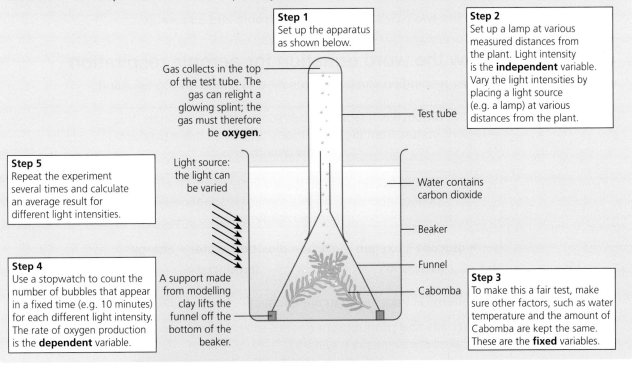

Step 1
Set up the apparatus as shown below.

Step 2
Set up a lamp at various measured distances from the plant. Light intensity is the **independent** variable. Vary the light intensities by placing a light source (e.g. a lamp) at various distances from the plant.

Gas collects in the top of the test tube. The gas can relight a glowing splint; the gas must therefore be **oxygen**.

Test tube

Step 5
Repeat the experiment several times and calculate an average result for different light intensities.

Light source: the light can be varied

Water contains carbon dioxide

Beaker

Funnel

Step 4
Use a stopwatch to count the number of bubbles that appear in a fixed time (e.g. 10 minutes) for each different light intensity. The rate of oxygen production is the **dependent** variable.

A support made from modelling clay lifts the funnel off the bottom of the beaker.

Cabomba

Step 3
To make this a fair test, make sure other factors, such as water temperature and the amount of Cabomba are kept the same. These are the **fixed** variables.

Exam-style questions: Exercise 7B

1 Choose words from the list to complete the following sentences.

**biomass   light   roots   glucose   water   chlorophyll   chemical
carbon dioxide   oxygen   green   Sun**

a) During photosynthesis, plants use _____ from the air and _____ from the soil to make _____ . (3)

b) The energy needed for photosynthesis is transferred from the nuclear store of energy in the _____ by _____ to a _____ store of energy in the plant. (3)

c) The energy is transferred to a _____ pigment called _____ . (2)

d) Photosynthesis provides _____ , which is needed for aerobic respiration. (1)

e) _____ absorb water and minerals from the soil. (1)

f) Plant growth causes an increase in _____ . (1)

2 Your class has set up an experiment to test that starch is produced by plants.

a) What substance do you use to test for the presence of starch? (1)

b) Describe the colour change that takes place when starch is present. (1)

c) Explain why you generally test for starch, rather than glucose. (1)

3 Explain why photosynthesis is important to the atmosphere. (1)

4 Describe the role plants play in food chains. (1)

Ask yourself

What would happen if there was no light on Earth? Think through the consequences.

8 Respiration

Respiration is a series of chemical reactions carried out within each living cell to **release energy** from chemical energy stores for all life processes. All living organisms respire.

There are two types of respiration: **aerobic** and **anaerobic**.

Know the word equation for aerobic respiration

Aerobic respiration can be summarised by the following equation:

> **Glucose** is a store of chemical energy. In animals, this comes from digested foods. In plants, it is a product of photosynthesis.

> **Oxygen** comes from the air inhaled into the lungs during breathing. So this form of respiration is called aerobic respiration.

glucose + oxygen → carbon dioxide + water + energy*

(reactants)   (products)

*The energy is transferred to the chemical stores of energy in the body.

All reactants and products are carried to and from the cells in the bloodstream (also called the circulatory system).

Oxygen enters the bloodstream, and carbon dioxide leaves it, by passing through the membranes of air sacs in the lungs. This is known as gas exchange.

Know the word equations for anaerobic respiration

Anaerobic respiration is when no oxygen is involved in respiration.

In humans (and animals)

Sometimes the blood cannot deliver oxygen to cells fast enough. For example, during strenuous exercise. If no energy is available, the body cannot function. Anaerobic respiration acts as a temporary backup system, allowing the stored energy in glucose to be transferred to energy stores in the body **without oxygen**.

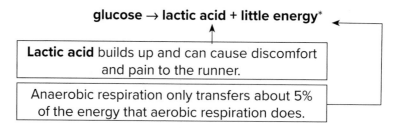

glucose → lactic acid + little energy*

> **Lactic acid** builds up and can cause discomfort and pain to the runner.

> Anaerobic respiration only transfers about 5% of the energy that aerobic respiration does.

*The energy is transferred to the chemical stores of energy in the body.

In yeast

In plants and yeast, glucose is converted to:

glucose → carbon dioxide + ethanol + little energy*

*The energy is transferred to the chemical stores of energy in the yeast.

Understand the importance of respiration, the release of energy from food, to carry out life processes

Respiration occurs continuously in living cells. Aerobic respiration is a series of chemical reactions involving glucose and oxygen that takes place in and around the mitochondria. As these reactions happen, energy in the chemical store of glucose is transferred to other energy stores in the body. Energy is needed for life processes to continue. Without it, the organism will die.

Recognise the importance of fermentation in yeast to human society

Anaerobic respiration in yeast is called **fermentation** and is at the heart of making bread and many drinks.

The holes in bread are caused by the production of carbon dioxide during the anaerobic respiration of the yeast – the bread is said to have 'risen'. During baking, the ethanol produced evaporates from the bread.

Carbon dioxide can also be used in fizzy drinks. The ethanol produces alcoholic drinks.

Recommended practical activities

1. **Investigate the exhalation of carbon dioxide during gas exchange in humans.**

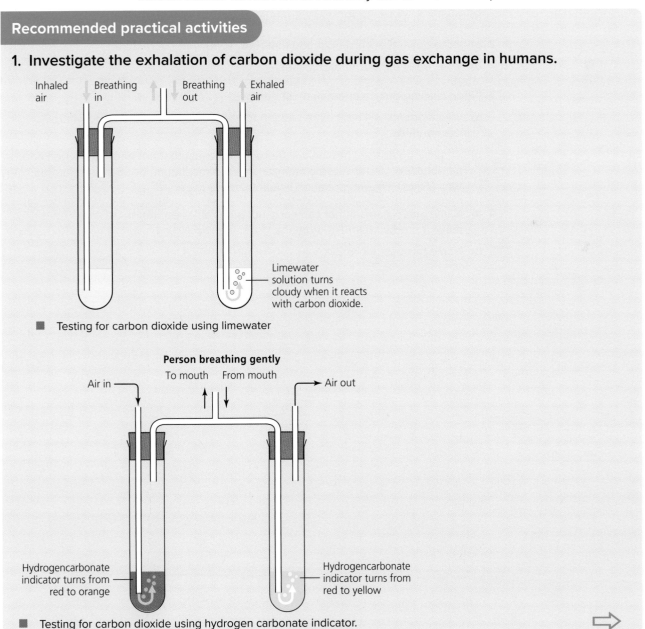

Inhaled air | Breathing in | Breathing out | Exhaled air

Limewater solution turns cloudy when it reacts with carbon dioxide.

■ Testing for carbon dioxide using limewater

Person breathing gently

Air in | To mouth | From mouth | Air out

Hydrogencarbonate indicator turns from red to orange

Hydrogencarbonate indicator turns from red to yellow

■ Testing for carbon dioxide using hydrogen carbonate indicator.

2. Investigate fermentation in yeast.

Step 1   Dissolve sugar in warm water

Step 2   Add yeast and pour suspension into a boiling tube

Step 3   Add a layer of oil

Step 4   Connect to a second tub containing limewater or hydrogencarbonate indicator to collect gas being released.

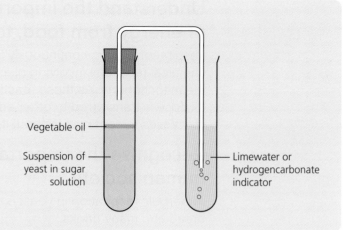

Vegetable oil

Suspension of yeast in sugar solution

Limewater or hydrogencarbonate indicator

Exam-style questions: Exercise 8

1 a) What does the word respiration describe?   (2)

   b) Where in an organism, does aerobic respiration take place?   (1)

2 a) Write down the word equation that represents aerobic respiration.   (2)

   b) What does the term aerobic mean?   (1)

3 When chasing prey, a cheetah must run fast.

   a) Which organs will help extra oxygen to reach the parts of the body where it is needed.   (2)

   b) At some point, the cheetah will be running so fast that not enough oxygen can be obtained. Under these circumstances the respiration becomes anaerobic. Why is this form of respiration valuable to the cheetah?   (1)

4 a) Name one organism that can carry out anaerobic respiration to make a product useful to humans.   (1)

   b) Name this useful product.   (1)

Preliminary knowledge

- All organisms need to reproduce. Chapter 6: Reproduction in humans, compares sexual and asexual reproduction in plants with sexual reproduction in animals.
- Fertilisation is the fusing of male and female sex cells in sexual reproduction.
- In plants, the flower contains the reproductive organs and its main function is to make seeds. Each seed that germinates successfully will grow into a new plant.
- Some flowers rely upon being visited by insects to help with reproduction. In order to attract insects, the flower will put on an attractive display (colour and shape of petals), or produce a distinctive smell (scent), or both.

Know the similarities and differences between insect- and wind-pollinated flowers

- The variety of plants depends upon the process of sexual reproduction, which is the fusion of the male and female sex cells that are made in sex organs. The process of carrying male pollen to a female stigma is called **pollination**.
- **Fertilisation** is the fusion of the **male gamete** (a nucleus in the pollen grain) with the **female gamete** (a nucleus in the ovule). Pollen grains and ovules are sex cells because they contain male and female gametes, but the pollen grain does not fuse with the ovule. The pollen grain stays on the **stigma** of the flower and sends out a pollen tube that grows through the style towards the ovule. This pollen tube contains the male gamete (nucleus).
- **Pollination** is the transfer of pollen from **anther** to stigma. Pollen can be carried from anther to stigma by either insects or the wind.

Study the following two diagrams to see the similarities and differences between insect- and wind-pollinated flowers.

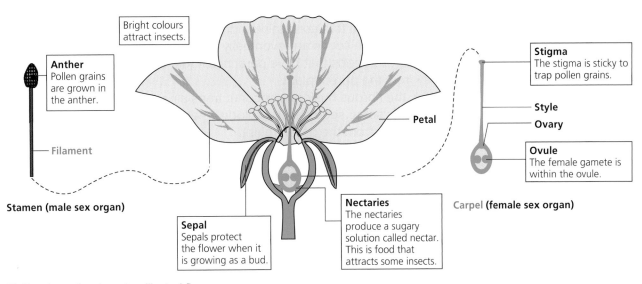

■ Structure of an insect-pollinated flower

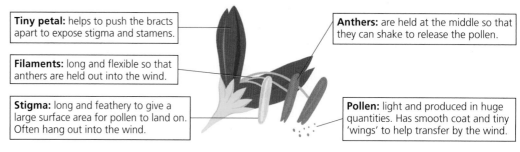

Tiny petal: helps to push the bracts apart to expose stigma and stamens.

Filaments: long and flexible so that anthers are held out into the wind.

Stigma: long and feathery to give a large surface area for pollen to land on. Often hang out into the wind.

Anthers: are held at the middle so that they can shake to release the pollen.

Pollen: light and produced in huge quantities. Has smooth coat and tiny 'wings' to help transfer by the wind.

■ Structure of a wind-pollinated flower

Understand the role of pollen grains in transporting male gametes to the female egg cell

The pollen grain carries the male sex cell (gamete) to the stigma. What follows is shown in the diagram below:

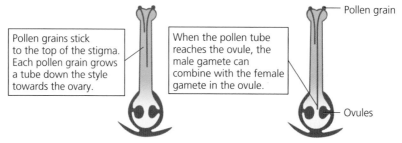

Pollen grains stick to the top of the stigma. Each pollen grain grows a tube down the style towards the ovary.

When the pollen tube reaches the ovule, the male gamete can combine with the female gamete in the ovule.

Pollen grain

Ovules

■ Pollen grain transporting the male gamete to the female gamete

The fertilised ovule then develops into a seed.

Recognise that different seed dispersal mechanisms result in new plants growing away from the parent plants

● Seeds are dispersed so that new plants do not compete with the parent plant for nutrients, light, water and space.
● Seeds can be dispersed in various ways. For example:
    ● Animals eat ripe fruit containing indigestible seeds. The animal moves around and the seeds are eventually excreted in some other place.
    ● Wind carries seeds that have parachutes (dandelion) or wings (sycamore) away from the parent plant.
    ● Water: the fibrous husk of a coconut traps air and helps the fruit to float.
    ● Explosion: some seed cases explode when they dry out, throwing the seeds away from the parent plant.
● The seeds germinate into new plants.

Recommended practical activity

Observe, interpret and record the similarities and differences between a wind-pollinated flower and an insect-pollinated flower.

Examples of wind pollinated plants:

- Dandelion seeds
- The seeds of a cottonwood tree being dispersed by the wind
- The wings on maple seeds help them travel on the wind

Examples of insect pollinated plants:

- Bluebells
- Honeysuckle
- Daisy

Exam-style questions: Exercise 9

1 Select the best word from the list to complete the sentences below.

   anthers   fertilisation   ovules   apples   fruits   pollination
   buds   germinate   stigmas   change   grow   testes

  a) In flowering plants, pollen is produced by the _____ . (1)

  b) The ovaries of flowering plants contain _____ . (1)

  c) The transfer of pollen between plants is called _____ . (1)

  d) After _____ , seeds are formed in the ovaries. (1)

  e) The ovaries, containing seeds, will develop into _____ . (1)

2 **a)** Some flowers are pollinated by insects. Name **one** feature that will help to attract insects. (1)

  b) Some seeds are dispersed by the wind. Name **one** feature of the seed that makes this an effective method of dispersal. (1)

  c) Some seeds are dispersed by animals. Name **one** feature of the seed that makes this an effective method of dispersal. (1)

10 Recreational drugs and human health

A healthy lifestyle depends on three main factors:

- a healthy diet
- taking regular exercise
- avoiding the intake of harmful substances.

Know that many substances, including alcohol, can affect the brain, the nervous system and our health

Our bodies are at risk from the following:

- Smoking can lead to lung cancer, difficulty breathing, and problems in the circulatory system and heart.
- Alcohol slows reactions. Excess amounts cause damage to the liver, stomach and heart.
- Drugs other than those prescribed for medical reasons (for example, solvents, cannabis, cocaine and amphetamines) can:
  - introduce chemicals into our bodies that upset the finely balanced chemical mechanisms that exist to keep all our life processes working properly
  - cause damage to the brain, heart, liver and stomach
  - lead to exposure of diseases through needles used to inject drugs (HIV, hepatitis).

Know how behaviour is affected

In addition to the effects on health, drugs can affect behaviour.

- In small amounts, alcohol can lower social inhibition and lead to pleasant feelings.
- In larger amounts, alcohol can increase the risk of accidents and injury, increase risk-taking behaviour, aggression and violence.
- Some drugs (such as marijuana) can make people lose touch with reality, miss deadlines, see things that are not there and become paranoid, feeling threatened or watched. They may also feel unhappy, with a low mood (depression).

Know and recognise the potential for addiction to certain medical and recreational drugs

Addiction is when you have no control over your actions as a result of taking, doing or using something. Often you cannot function without the drug.

- Smoking, some drugs and alcohol can become addictive.
- Addiction can cause very serious long-term changes to behaviour.

Understand the importance of cleanliness at personal and community levels as a defence against disease

Fighting disease

Diseases are either non-infectious (not caught from somebody else) or infectious (caught from somebody else).

Causes of infectious diseases

Micro-organisms (microbes) divide and reproduce very rapidly and can damage cells or release toxins (poisons) that can make people feel very ill.

Viruses:

- must invade a living cell to reproduce, eventually causing damage to the cell
- when they are in a host cell, cannot be destroyed without damaging the cell
- cannot be controlled by antibiotics.
- COVID-19, Ebola, and influenza are examples of diseases caused by viruses.

Bacteria:

- live and grow outside living cells
- can reproduce every 20 minutes
- are bigger than viruses, but smaller than cells
- can be killed by:
  - antibiotics – taken as medicine
  - antiseptics – on the skin
  - disinfectants – on kitchen and bathroom surfaces.
- Cholera, tetanus and various types of food poisoning are caused by bacteria.

Natural defences

The human body has three natural defences against disease: barriers, white blood cells and blood clots.

- Skin, wax in ears and tears in eyes all form barriers to keep microbes away from the body's tissues.
- Open wounds are protected from microbes by blood clots, which are produced by a special type of blood cell (platelet), forming a scab.

Action humans can take as defences against disease

Personal hygiene

- Regular washing of hands, hair and the body removes bacteria, which cause body odours, infestations in hair and spread of food poisoning.
- Regular brushing of teeth removes bacteria that cause tooth decay.

Community actions

- Provision of safe, clean drinking water.
- Removal and safe disposal of refuse and sewage.
- Provision of medical care:
  - immunisation
  - medicines such as antibiotics.

Recommended practical activity

Make a presentation showing the risks of common medical and recreational drugs.

Exam-style questions: Exercise 10

1 Match the following features of a lifestyle with the problems they cause. (4)

Lifestyle	Problem
A Smoking	**1** may lead to blood diseases such as HIV and hepatitis.
B Alcohol	**2** causes lung cancer.
C Injecting drugs	**3** causes long-term changes to behaviour.
D Addiction	**4** damages the liver.

2 a) Name **two** ways in which a community can reduce the spread of infectious diseases. (2)

b) Give **one** example of an infectious disease. (1)

c) What medicine could a doctor prescribe to treat a bacterial infection? (1)

3 If a bacterium could divide into two every 20 minutes, how many bacteria would there be after 3 hours? (2)

The interdependence of organisms in an ecosystem

Preliminary knowledge

Food chains describe the relationship between the types of organism in an ecosystem by describing what each organism eats and what eats them.

Food chains show how energy is transferred between the chemical stores of energy in organisms.

Feeding relationships (food chains)

A → B means that organism A is eaten by organism B. The arrow points to the direction that energy and nutrients are transferred through the food chain. For example:

rose (leaves) → aphid → ladybird → robin

producer → **consumer** → **consumer** → **consumer**

In this chain:

plant → herbivore → carnivore → top carnivore

- Food chains usually start with green plants. Green plants can make their own food by photosynthesis, so these are called **producers**.
- Animals that eat other organisms are called **consumers**.
- Animals that eat only plants are called **herbivores**.
- Animals that eat other animals are called **carnivores**.
- Animals that eat plants and animals are called **omnivores**.
- **Predators** are animals that eat other animals (their **prey**).

Test your preliminary knowledge: Exercise 11A

1 Four food chains are given below:
A lettuce → rabbit → fox
B oak tree → aphid → ladybird → robin
C grass → earthworm → shrew → owl
D algae → pond snail → leech → dragonfly nymph
  a) Which food chain occurs in water? (1)
  b) Which food chain does not contain a vertebrate? (1)
  c) In which food chain is the herbivore a mammal? (1)
  d) Which food chains do not contain an insect? (1)
  e) In which food chain is the producer much larger than the herbivore? (1)

Know how food webs give information about producers and consumers within an ecosystem

Without green plants, life as we know it would not exist at all.

- There would be no oxygen in the air.
- There would be no food – no vegetables, salads or fruit; no meat from animals that have fed on plants.

Remember these key definitions:

Habitat

All organisms need a place to live that will provide them with food, shelter and protection from predators. The place where a living organism lives (for example, a pond, a hedgerow, a wood) is called a **habitat**.

Community

- This refers to all of the living organisms within a habitat.
- The **community** consists of a collection of organisms of different species.
- A habitat (a place) and the communities (collection of living organisms), add together to make an **ecosystem**.

Food chains

Food chains show the feeding relationships within a habitat, how energy is transferred between the chemical stores of energy in one organism in the chain to another.

Food webs

Food webs are interconnected food chains. Many have more than one supply of food. Feeding relationships are complex and food chains overlap. Here is an example of a food web showing the relationships between producers and consumers. Food webs give us a more realistic picture of the feeding relationships in a habitat.

TOP CARNIVORES: Foxes and buzzards are at the top of this food web because nothing else eats them. There can only be a few top carnivores because each one of them needs to consume many organisms 'lower down' in the web.

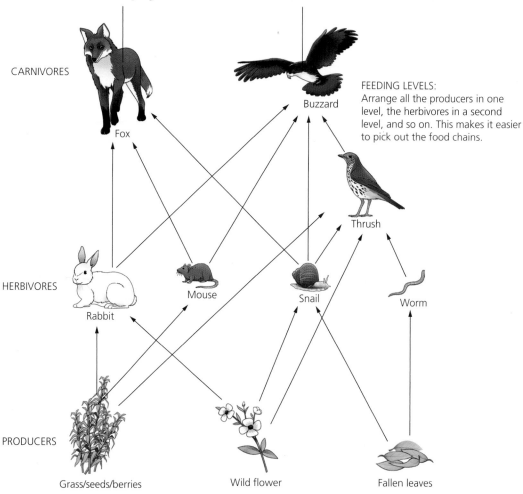

CARNIVORES

Fox

Buzzard

FEEDING LEVELS:
Arrange all the producers in one level, the herbivores in a second level, and so on. This makes it easier to pick out the food chains.

Thrush

HERBIVORES

Rabbit

Mouse

Snail

Worm

PRODUCERS

Grass/seeds/berries

Wild flower

Fallen leaves

Any disturbance to the numbers of organisms at one level will create an imbalance. For example, the removal of the top carnivores would allow herbivore numbers to increase. This in turn would mean more plants are eaten and there are, therefore, fewer producers to make food by photosynthesis.

Understand the impact of human activities on food webs

Humans are omnivores and dominant consumers, eating animals and plants. As the world population increases, so too does the demand for food. This demand for food can have significant effects on the world's ecosystems and the food webs within them. The changes and their impact include:

- **Habitat destruction** Agriculture, deforestation, poisoning of water supplies due to the use of fertilisers, extensive demand for water, over fishing.
- **Global warming and climate change** Use of fossil fuels emitting carbon dioxide and other greenhouse gases is leading to rising temperatures that are altering ecosystems on land (for example, forests) and in the ocean (for example, reefs).
- **Overpopulation** Death rates have decreased and medicines are keeping people alive for longer.
- **Waste from industry, farms and homes** Poisonous chemicals released into the water, animals eating plastic and dying.
- **Invasive species** being transported round the world are destroying native species (for example, zebra mussels from Russia are causing problems in the Great lakes between the USA and Canada).

Understand the need for conservation and sustainable development

Research into food webs identifies the links between producers and consumers, and consumers and their predators, and the vulnerability of those ecosystems to species loss. It can be a powerful guide to what needs to be done to protect them.

Conservation – human-made ways of helping the environment

- National parks, wildlife centres, zoos.
- Protection of endangered species and their habitats.
- Reforestation.
- Vehicles that are more efficient to reduce harmful emissions produced when burning fossil fuels.
- Greater use of alternative forms of energy (wind, geothermal, solar) to reduce pollution from burning fossil fuels.
- Recycling of household waste to reduce landfill.

Sustainable development

- It is important to manage what we take from the environment now to protect it for the future generations.
- Conservationists try to balance human demands on the environment with the need to maintain wildlife habitats.
- Some examples: using solar and wind energy, creating green spaces in cities, crop rotation, sustainable forestry, water conservation.

Recognise the interdependence of organisms in food webs

Food webs show how food chains overlap. If an organism within a food web is affected in some way, it can have an impact on many food chains. Damage to these ecosystems can cause the whole food web to disappear.

Recommended practical activities

Investigate producers and consumers in a local habitat, using:

1. Quadrats

A **quadrat.** Place this in the habitat.

Count the number of organisms of a particular species inside the quadrat.

Work out how many quadrats fit into the habitat.
Now calculate how many organisms are in the whole of the habitat.

Rock pool habitat.

■ Measuring population using a quadrat to count living organisms in a habitat

2. Pitfall traps

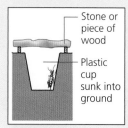

- Stone or piece of wood
- Plastic cup sunk into ground

■ Measuring population using a pitfall trap to catch small animals or insects

3. Other sampling techniques

■ A net can be used to catch swimming organisms

■ A net can be used to catch flying organisms

Exam-style questions: Exercise 11B

1 Buzzards (large birds of prey) were almost hunted out of existence, but they have now made a huge comeback, with numbers increasing in the past few years. Give three effects this will have on the food web from earlier in this chapter. (3)

2 List three ways in which human activities could affect food webs. (3)

3 List three ways in which conservation of the environment can be carried out. (3)

Ask yourself

What the actions could you take to protect or improve your environment?

Preliminary knowledge

The variety of plants and animals makes it important to identify them and assign them to groups.

Diagnostic features

The differences between living organisms within a species are called **variations**. We can use the **diagnostic features** of organisms to produce identification keys.

Spider or branching key

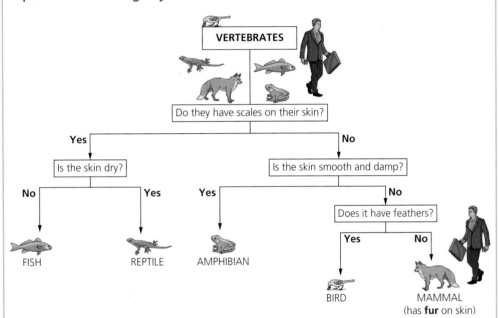

Number key

1	Do they have scales on their skin?
	YES – Go to question **2**
	NO – Go to question **3**
2	Is the skin dry?
	YES – Reptile
	NO – Fish
3	Is the skin smooth and damp?
	YES – Amphibian
	NO – Go to question **4**
4	Does it have feathers?
	YES – Bird
	NO – Mammal

Remember that some features of animals and plants are diagnostic (can help you assign an organism to a group) and some are not: for example, the type of skin is diagnostic in vertebrates whereas size and mass are not.

Classification

Sorting organisms into groups is called classification. Organisms are given Latin names, which identify them within particular groups, largely as a result of the pioneering work carried out in this field by Carl Linnaeus.

The largest groups are five **kingdoms** which include the plant and animal kingdoms. Kingdoms are divided into **groups**.

The animal kingdom

Animals fall into two main groups:

- Vertebrates – animals with a backbone and an internal skeleton, made of bone.
- **Invertebrates** – animals that do not have a backbone or a skeleton made of bone.

Arthropods are one of the sub-groups of invertebrates. All animals in this group have:

- jointed limbs (*arthro* = jointed; *poda* = limbs)
- a hard outer body covering
- segmented bodies.

This group is divided into **classes**. Two of these classes are insects and arachnids.

Class	Number of body parts	Number of legs	Antennae?	Wings?	Typical examples
Insects	3	6	Yes	Yes	bee, fly, beetle
Arachnids	2	8	No	No	tarantula, scorpion, mite, tick

The plant kingdom

All plants within this kingdom make their own food by the process of photosynthesis.

There are four main groups of plants within the plant kingdom: mosses, ferns, conifers and flowering plants (those that produce seeds).

Fungi are in a separate kingdom because they do not contain chlorophyll, do not carry out photosynthesis and obtain their food from their surroundings.

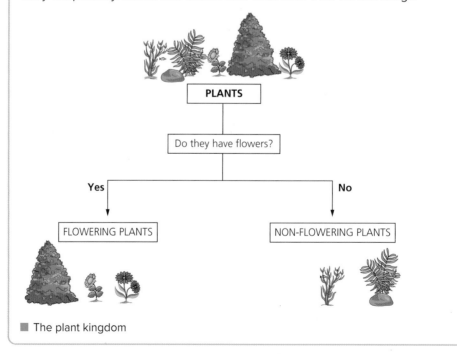

■ The plant kingdom

Test your preliminary knowledge: Exercise 12A

1 Draw a branching key to display the information about vertebrates given below:

  1 Have scales on their skin   Go to **2**
     No scales   Go to **3**
  2 Dry skin   reptile
     Moist skin   fish
  3 Smooth and damp skin   amphibian
     Rough and dry skin   Go to **4**
  4 Has feathers   bird
     No feathers (fur on skin)   mammal   (5)

2 'Come into my parlour said the spider to the fly.' Explain what these two animals have:

  a) in common   (1)
  b) as differences between them.   (3)

3 Make a table with two headings, 'invertebrate' and 'vertebrate'.

  a) Put the subheadings 'backbone', 'no backbone' into the correct table heading.   (1)
  b) Add the following animals to the correct column of your table:
  octopus   **cat**   **shark**   **spider**   **frog**   **crab**   **beetle**   **fox**   (8)

4 A lizard is a reptile and a newt is an amphibian. In many ways, they look similar, yet they belong to different classes. Give reasons for this.   (2)

Know how living organisms are classified into five kingdoms

The largest groups of organisms are known as kingdoms. There are five kingdoms.

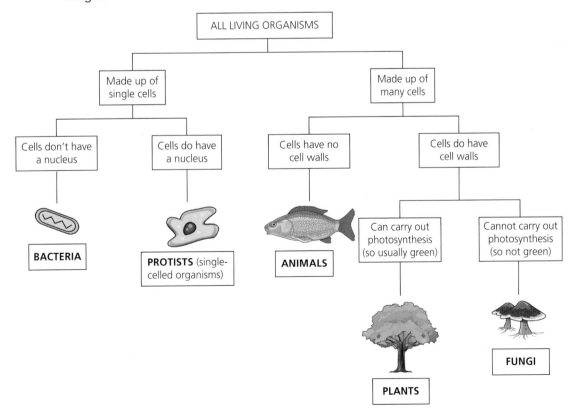

■ The five kingdoms of living organisms

Recognise the similarities and differences between the structures and life processes of organisms in the different kingdoms

The animal kingdom

All animals can be put into one of two groups: invertebrates and vertebrates. They can be put into the correct group by asking one question: Do they have a backbone?

Invertebrates

Invertebrates have no backbone. The main groups within a kingdom are called **phyla** (the singular is phylum).

Arthropoda is an important invertebrate phylum. All animals in this phylum have:

- jointed limbs in pairs (*arthro* = jointed; *poda* = limbs)
- a hard outer covering (exoskeleton) made of chitin or limestone
- bodies divided into segments (compartments).

This is the largest of all animal phyla and is divided into four main **classes**:

- Insects – bees, flies, beetles, and so on
- Arachnids – including spiders, scorpions, mites and ticks
- Crustaceans – which includes not only sea creatures, such as prawns, lobsters and crabs, but also land animals such as woodlice
- Myriapods – centipedes and millipedes

Remind yourself of the differences in the structures of insects and arachnids (see table earlier in this chapter).

Vertebrates

Vertebrates have a backbone made of interlocking bones called **vertebrae**. These protect the nerve cord, which is connected to the brain. The brain is located in the skull and so is itself protected by bone.

Vertebrates are members of a single phylum in which there are five main **classes**. You can generally identify the five groups by their skin, as in the keys earlier in this chapter. There are other important differences:

	Reptiles	Fish	Amphibians	Birds	Mammals
Constant body temperature?	No: they are cold-blooded	No: they are cold-blooded	No: they are cold-blooded	Yes: they are warm-blooded	Yes: they are warm-blooded
Do they lay eggs?	Yes – on land; eggs have soft shells	Yes – in water	Yes – in water	Yes – on land; eggs have hard shells	No
Do they feed milk to their young?	No	No	No	No	Yes
Where do they live?	On land and/or in water	In water	On land and/or in water	In air or/and on land or/and in water	In air or/and on land or/and in water
Examples	snake, crocodile, turtle	haddock, cod, shark	frog, toad, newt	eagle, blackbird, emu	dog, elephant, whale, human, bat

- **Cold-blooded animals** – body temperature changes according to the temperature of the surrounding air or water.
- **Warm-blooded animals** – keep a constant body temperature.

Recognise the diagnostic features of arthropods

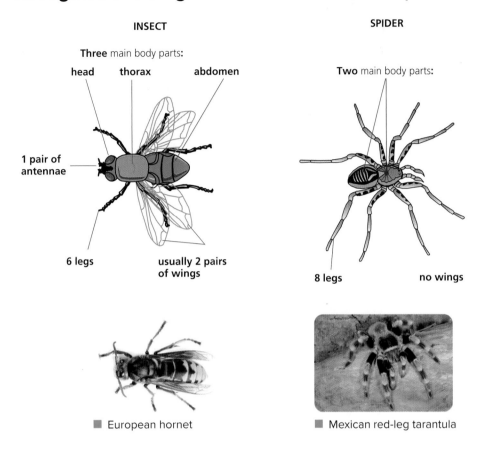

INSECT

Three main body parts:

head   thorax   abdomen

1 pair of antennae

6 legs   usually 2 pairs of wings

SPIDER

Two main body parts:

8 legs   no wings

■ European hornet

■ Mexican red-leg tarantula

Recognise the diagnostic features of vertebrate groups

Revisit and make sure you know the differences between fish, amphibians, reptiles, birds and mammals.

The plant kingdom

All plants have one thing in common; they have a pigment (chlorophyll) that can absorb light for photosynthesis.

Photosynthesis transfers energy from the nuclear store of the Sun to a chemical store of energy in the plant.

The plant kingdom has four main groups shown in the table below.

Feature	Mosses	Ferns	Conifers	Flowering plants
Leaves	Simple	Yes	Yes	Yes
Stem	No	Yes	Yes	Yes
Root	Yes	Yes	Yes	Yes
Flower	No	No	No	Yes

Recommended practical activities

1. Observe the structures of various organisms, relating structure to function. For example:

■ Bird feathers are used for flight

■ Conifer cones are related to seed dispersal

■ The gills of fish are used to obtain oxygen (for respiration) from water

2. Observe and interpret photographs, specimens and videos to compare diagnostic features of key groups of organisms.

Exam-style questions: Exercise 12B

1 Match each organism in the first column with the correct description in the second column. (9)

Organism		Description	
A	Spider	1	Made up of many cells, cells have cell walls, does not photosynthesise
B	Bluebottle (fly)	2	Cells have cell walls, photosynthesises, has leaves, roots and stem but no flowers
C	Fern	3	No backbone, 2 body parts, 8 legs, no wings
D	Toadstool	4	Cold-blooded, lays eggs with soft shells, has a backbone, dry skin
E	Daisy	5	No backbone, 3 body parts, 6 legs, wings
F	Frog	6	Many cells, cells have walls, has flowers for reproduction
G	Crocodile	7	Has hair, constant body temperature, feeds young with milk
H	Cat	8	Single cell and no nucleus
I	Bacterium	9	Has a backbone, cold-blooded, lays eggs in water

2 Grass, rabbits and mushrooms are all organisms found in a wood, and yet they are all placed in different kingdoms. Explain why. (3)

Variations in living organisms

Although humans belong to the same species of animal (*Homo sapiens*), there are clearly differences between individuals – these are called **variations**.

Know the distinction between discontinuous and continuous variations

Discontinuous variations

- Discontinuous variations can be put into groups easily.
- Discontinuous variations enable easy sorting of organisms; there are no in-between groups and they depend only on genes. For example, you can only have blood group A, B, AB or O.

Continuous variations

- Continuous variations fall into many groups that almost run into each other. Height, shape, weight and build are variations that result from **inherited genes** working with their **environment**. Clearly, the amount of food intake and exercise will have an effect on all of these variations, as well as those caused by the passing on of genes.
- Continuous variations are not easy to put into discrete groups as there are many groups for each particular feature. For example, with height, any individual will have a measurement that can fit anywhere in a continuous distribution from the minimum to the maximum.

Recording and presenting data

Remember how you can record this data.

Discontinuous data – genes only

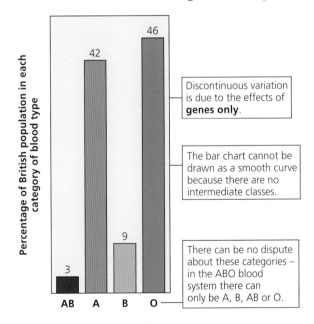

■ Human blood groups are an example of discontinuous variation. You can only be in one of four distinct blood groups – AB, A, B or O – there are no in-between classes.

Continuous data – genes working with their environment

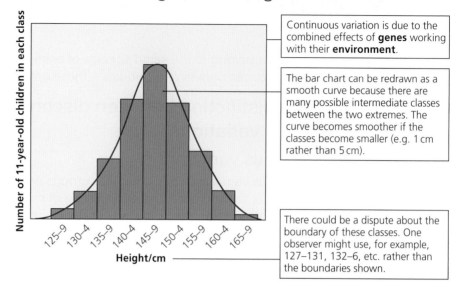

Continuous variation is due to the combined effects of **genes** working with their **environment**.

The bar chart can be redrawn as a smooth curve because there are many possible intermediate classes between the two extremes. The curve becomes smoother if the classes become smaller (e.g. 1 cm rather than 5 cm).

There could be a dispute about the boundary of these classes. One observer might use, for example, 127–131, 132–6, etc. rather than the boundaries shown.

■ Human height is an example of continuous variation. Other examples include quantitative characteristics such as chest circumference, body mass (weight) and hand span.

Recommended practical activities

Revisit any practical activities you may have done related to variation. For example: measuring and comparing biometric data of living organisms such as members of your class (height, eye colour and hand span).

Exam-style questions: Exercise 13

1 a) To which type of variation do body mass and ability belong? (1)

 b) From what do these result? (1)

2 a) Which of these are examples of discontinuous variation? (1)

 height   eye colour   length of hair   shoe size   blood group

 b) Explain why you made the choice you did. (1)

 c) Explain why you rejected the others. (1)

Biology – Test yourself

Before moving on to the next chapter, make sure you can answer the following questions. The answers are at the back of the book.

1 Draw diagrams of typical animal and plant cells and include labels for the following: cell surface membrane, nucleus, cytoplasm, cell wall, large vacuole, chloroplasts, starch grains for storage and mitochondria.

2 What do the following words describe: tissue, organ, system?

3 Humans need a diet containing carbohydrates, fat, minerals, protein and vitamins.
 a) i) In a healthy diet, which one of these gives us most of our energy?
   ii) Name a food rich in this type of substance.
 b) i) Which of these is needed for the growth and repair of cells?
   ii) Name a food rich in this type of substance.

4 Complete the following sentences about reproduction in humans.
 a) The male and female sex organs mature during a stage of development called ____ .
 b) The male gamete is the ____ and this is produced in the ____ .
 c) The female gamete is the ____ and this is produced in the ____ .
 d) Male and female gametes fuse together during fertilisation to form a ____ .
 e) ____ is the process of the embryo settling into the thickened lining of the uterus wall.
 f) Once the process in e) has been completed, the mother is ____ .

5 List two ways in which regular exercise is an essential part of healthy living.

6 Explain what photosynthesis is.

7 What are the raw materials needed for photosynthesis?

8 What are the products of photosynthesis?

9 What happens to the oxygen produced by photosynthesis?

10 What happens to the glucose produced by photosynthesis?

11 Nitrogen is present in all proteins; magnesium is needed to make chlorophyll.
 a) Where will plants find supplies of these chemicals?
 b) Which structures enable these chemicals to enter the plants?

12 Explain what the word variation means.

13 What does a scientist mean by the word classification?

14 One man and his dog have much in common, even though they are clearly different species. Say what they have in common in terms of biological classification.

15 Name three features that are common to all arthropods.

16 Make a table with two headings, 'vertebrate' and 'invertebrate'. Put the following animals in the correct part of your table.

| earthworm | emu | shark | spider | whale |
| frog | crab | turtle | beetle | fox |

17 What is an ecosystem?

18 What does the term environment mean?

19 Here is a simple food chain:

rose → aphid → robin → cat

 a) Which organism is the primary producer and why is it given this name?
 b) Suggest two effects on the food chain of spraying the roses with insecticide.

20 a) What is a food web?
 b) Explain how a food web is different from a food chain.

Preliminary knowledge

The particle model

- All substances (matter) are made from tiny particles called **atoms**.
- There are about 100 different types of atom.
- Each **element** has its own type of atom.
- Atoms join together to form **molecules**.

States of matter

- Substances can be put into three main groups, known as the three states of matter: solid, liquid and gas.
- Most substances can exist in all three states; they can change from one to another by a change in temperature.
- There is no change in mass when a substance changes state.
- Particles move all the time.
  - Liquids and gases are called fluids because they can change shape and flow.
  - In a fluid, the movement is from one place to another. This is called diffusion (see below).
  - In a solid, the movement is in one place in the form of vibrations.

Particles in solids

- Particles are packed closely together (for example, regular patterns in crystals).
- They are held together strongly by various forces, so particles can only vibrate where they are and do not move around. This is why solids do not flow and keep their shape.
- Because particles are so close together, solids cannot be squashed into a smaller volume.

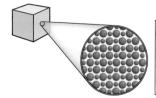

Particles in a solid are fixed in a pattern, so you cannot change its volume or easily change its shape.

Particles in liquids

- Particles are very close together, so a liquid cannot be squashed; the volume remains the same when you squeeze it.
- They are constantly moving around each other as they are not held together as strongly as in a solid, so a liquid is able to flow from one place to another.
- A liquid will change shape and match the shape of the container it is in, even though the amount of liquid (volume) remains the same if it is put into different containers.

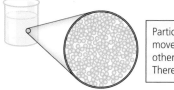

Particles in a liquid move around each other.
There is no pattern.

Particles in a gas

- Particles are relatively a long way away from each other. They move around rapidly in all directions, so a gas will completely fill any container.
- The big spaces between particles make it relatively easy to squash a gas into a smaller space (reduce its volume). This brings the particles closer together and it is sometimes possible to squash the particles so close together that the gas becomes a liquid.

⇨

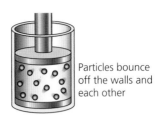
Particles bounce off the walls and each other

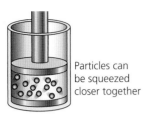
Particles can be squeezed closer together

A particular amount (mass) of a substance in each of the three states will show the following features (properties):

	Solid	Liquid	Gas
Mass	Fixed	Fixed	Fixed
Volume	Fixed	Fixed	Changes
Shape	Fixed	Changes	Changes

Changing between states of matter

Most substances can exist in all three states of matter; they can change between them.

- The process of changing a liquid into a gas is called **evaporation**.
- The process of changing a gas into a liquid is called **condensation**.

The **water cycle** is a good example of evaporation and condensation.

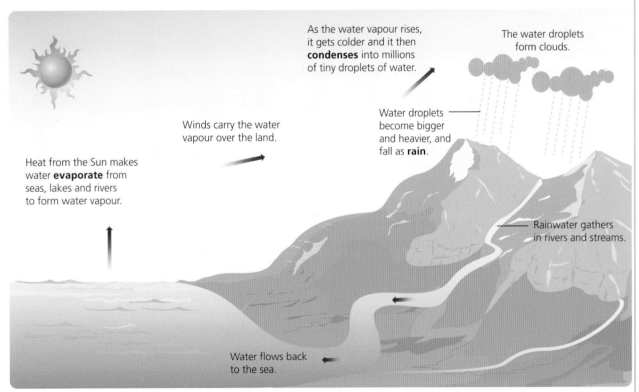

As the water vapour rises, it gets colder and it then **condenses** into millions of tiny droplets of water.

The water droplets form clouds.

Winds carry the water vapour over the land.

Water droplets become bigger and heavier, and fall as **rain**.

Heat from the Sun makes water **evaporate** from seas, lakes and rivers to form water vapour.

Rainwater gathers in rivers and streams.

Water flows back to the sea.

■ The water cycle: the processes of evaporation and condensation

The temperature at which a liquid changes into a solid is called the **melting point** (or freezing point).

- The freezing point of pure water is 0°C.
- The process of changing a solid into a liquid is called **melting**.
- The process of changing a liquid into a solid is called **freezing**.
- Both of these processes take place at the melting point of the substance.

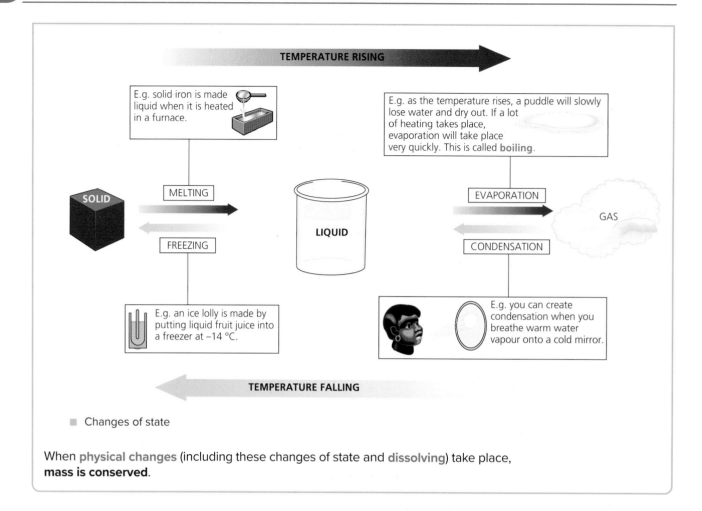

■ Changes of state

When **physical changes** (including these changes of state and **dissolving**) take place, **mass is conserved**.

Test your preliminary knowledge: Exercise 14A

1 There are three states of matter.

    a) Name these states.   **(1)**

    b) List **three** properties of each state.   **(3)**

2 Describe **i)** the position and **ii)** the movement of particles in:

    a) solids   **(2)**

    b) liquids   **(2)**

    c) gases   **(2)**

3 a) Name the two physical changes of state that take place in the water cycle.   **(2)**

    b) Complete these sentences.

        i) The process of changing liquid into a gas is called _____ .   **(1)**

        ii) The process of changing a gas into a liquid is called _____ .   **(1)**

4 Complete the following sentences.

    a) The process of changing a solid into a liquid is called _____ .   **(1)**

    b) The process of changing a liquid into a solid is called _____ .   **(1)**

Understand that changes of state are due to changes in motion and arrangement of particles

The three states of matter can be explained by the way the small particles (atoms or molecules) are arranged. Changes in state can be explained by particle theory.

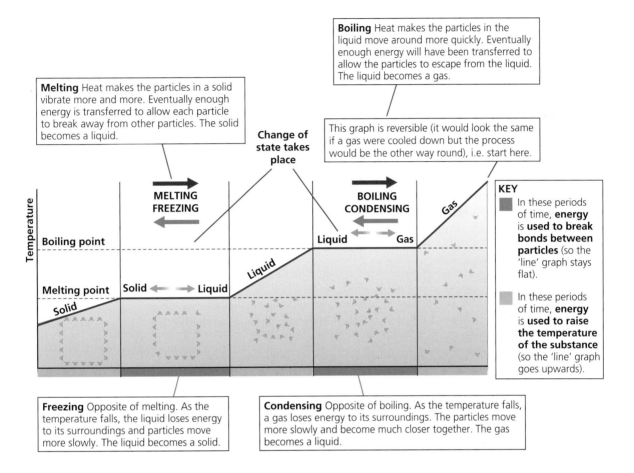

Boiling Heat makes the particles in the liquid move around more quickly. Eventually enough energy will have been transferred to allow the particles to escape from the liquid. The liquid becomes a gas.

Melting Heat makes the particles in a solid vibrate more and more. Eventually enough energy is transferred to allow each particle to break away from other particles. The solid becomes a liquid.

Change of state takes place

This graph is reversible (it would look the same if a gas were cooled down but the process would be the other way round), i.e. start here.

MELTING FREEZING

BOILING CONDENSING

Gas

Boiling point

Liquid — Gas

Liquid

Melting point — Solid ← → Liquid

Solid

KEY

In these periods of time, **energy is used to break bonds between particles** (so the 'line' graph stays flat).

In these periods of time, **energy is used to raise the temperature of the substance** (so the 'line' graph goes upwards).

Freezing Opposite of melting. As the temperature falls, the liquid loses energy to its surroundings and particles move more slowly. The liquid becomes a solid.

Condensing Opposite of boiling. As the temperature falls, a gas loses energy to its surroundings. The particles move more slowly and become much closer together. The gas becomes a liquid.

Sublimation

Some substances, including carbon dioxide and iodine, miss out the liquid state when they are heated or cooled. They change straight from solids to gases. This is called sublimation.

There are other important properties that help to define whether a substance is a solid, liquid or gas. Think about what is happening to the particles in these conditions.

	Solids	Liquids	Gases
Conduction (transfer of heat)	Solids that are **metals** are good **conductors**	Not good conductors, apart from metals	Not good conductors
Expansion (a physical change when heated)	Expand	Expand more than solids	Expand more than liquids
Contraction (a physical change when cooled)	Contract	Contract	Contract

Understand that pressure of a gas is caused by collisions with walls of a container

Particles in a gas move around quickly and randomly. As they bump into the walls of the container, they create a force. This is called gas pressure.

Recognise examples of diffusion and explain this process in terms of the random movement of particles

Diffusion is the random movement of particles in liquids and gases (fluids).

A botanist called Robert Brown noticed some pollen grains moving about in all directions in a jar of still water. As pollen grains cannot move on their own, he deduced that particles of water were moving randomly in all directions and were pushing the pollen grains.

You may have seen illuminated smoke particles being pushed around in still air, showing that gas particles move around at random.

- Random movement of particles in liquids and gases is called Brownian motion.
- Diffusion is a result of the Brownian motion and collisions between particles.
- The blue colour spreading through still water when a crystal of hydrated copper sulfate is dropped in it is an example of diffusion.

Recommended practical activities

Changes in state can be observed in the following practical activities:

- Changes in melting solids: for example water
- Boiling liquids: for example water
- Subliming solids: for example iodine
- Diffusion of gases: for example ammonia
- Diffusion of solids in solutions: for example $KMnO_4$

Exam-style questions: Exercise 14B

1 Match the following changes of states with the correct definition. (4)

Change of state	Definition in terms of particle theory
A Freezing	**1** Heat makes particles in a solid vibrate more and more. Eventually enough energy is transferred to allow particles to break away from other particles.
B Condensing	**2** Heat makes the particles in the liquid move around more quickly. Eventually enough energy will have been transferred to allow the particles to escape from the liquid.
C Melting	**3** A liquid becomes a solid as the temperature falls. It loses energy to its surroundings and particles move less and less.
D Boiling	**4** As the temperature falls, gas becomes a liquid. It loses energy to its surroundings. Particles move more slowly and get much closer together.

2 a) What is the name of the process occurring when a substance changes state from a solid to a gas without becoming a liquid? (1)

   b) Give an example of a substance that undergoes this process. (1)

3 Complete the following sentences.
   a) Particles in a gas move around quickly and randomly causing _____ . (1)

   b) Diffusion is the _____ movement of particles in liquids and gases. (1)

Atoms and elements

Remember that some materials are better electrical conductors than others:

- Metals and carbon (graphite) are conductors of electricity so, for example, copper is used for household wiring.
- Most other materials are **insulators** so, for example, plastic is used for plug covers.

Some materials are better thermal conductors than others:

- Metals are good thermal conductors and so are used for the base of pans, for example.
- Air is a good thermal insulator; examples of situations where trapped air is used for insulation in everyday life include: winter clothing, sleeping bags, expanded polystyrene for cups.

Understand the simple Dalton model

John Dalton proposed a theory that all matter was made of invisible atoms and that atoms could join together to form **molecules**.

Understand the idea of an element containing one type of atom

- A substance that is made of **only one type of atom** is called an **element**.
- All atoms within an element behave in the same way.
- Atoms in one element are different to atoms in other elements.

Know symbols for these elements

- Scientists have given each element a different chemical symbol.
- Often these are the first one or two letters of an element's name in English (for example, C for carbon) or another language (copper is Cu from the Latin word *cuprum*).

Make sure you know the symbols for these elements:

Name of element	Chemical symbol
hydrogen	H
carbon	C
nitrogen	N
oxygen	O
sulfur	S
magnesium	Mg
sodium	Na
chlorine	Cl
calcium	Ca
copper	Cu
iron	Fe
helium	He

Know there are about 100 elements in the Periodic Table

There are about 100 elements and they are all listed in the **Periodic Table**.

This means there are about 100 different types of atom.

The Periodic Table

- The diagram below shows a small selection of these elements.
- The horizontal rows are called periods.

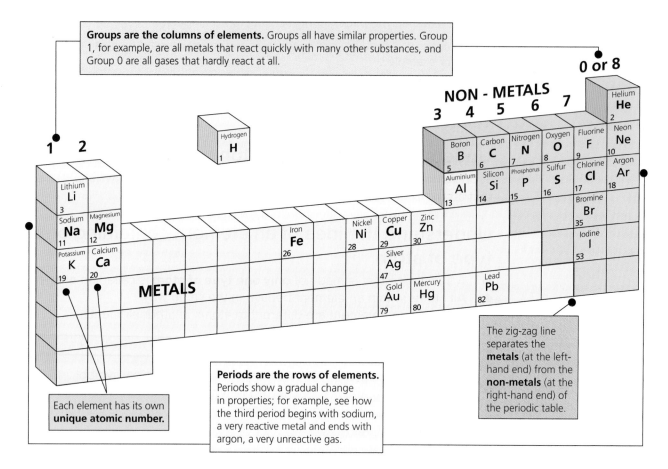

Groups are the columns of elements. Groups all have similar properties. Group 1, for example, are all metals that react quickly with many other substances, and Group 0 are all gases that hardly react at all.

Each element has its own **unique atomic number.**

Periods are the rows of elements. Periods show a gradual change in properties; for example, see how the third period begins with sodium, a very reactive metal and ends with argon, a very unreactive gas.

The zig-zag line separates the **metals** (at the left-hand end) from the **non-metals** (at the right-hand end) of the periodic table.

Recognise that a symbol stands for one atom of the element

- The chemical symbols used on the Periodic Table represent **one atom** of each of the elements.
- In a few elements the particles are molecules, when two or more **identical atoms** are bonded together.
- The **chemical formula** for an element indicates whether it is made of single atoms or of molecules.

Here are some elements and their chemical symbols and formulae:

Name of element	Symbol of atoms	Chemical formula of particles	Description of particles
helium	He	He	One He atom
oxygen	O	O_2	Molecule of two identical O atoms
sulfur	S	S_8	Molecule of eight identical S atoms

Know about types of element – metals and non-metals – their physical characteristics and differences in conductivity

Look at the Periodic Table and note the diagonal line towards the right that divides the table into two major groups: metals (on the left) and non-metals (on the right).

Metals and non-metals have different characteristics:

Metals	Non-metals
Solid at room temperature (except mercury – a liquid)	At room temperature: **solid:** carbon, iodine, sulfur **liquid:** bromine **gas:** oxygen, nitrogen, chlorine
Shiny	Dull
Bendy (malleable): can be bent, twisted or stretched	**Brittle:** snap or break into powder
Sonorous: make a bell-like sound when hit	Not sonorous
Usually **very dense**	Have a **low density**
Good conductors of heat and electricity	Poor conductors, so **good insulators** (except carbon which, in the form of graphite, conducts electricity and, in the form of diamond, conducts heat well)

Recommended practical activities

- Remember any investigations you have done to test for their conductivity. You may have tested iron, aluminium, sulfur, carbon in the form of graphite, and set up your apparatus like this:

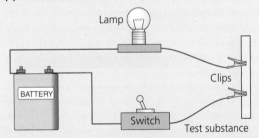

- Remember any investigations you have done to explore the hardness, appearance and other physical characteristics of a variety of elements.

Exam-style questions: Exercise 15

1. Complete the following sentences.
   a) A substance that is made of only _____ type of atom is called an element. (1)

   b) The chemical symbol for hydrogen is _____ . (1)

   c) The chemical symbol for chlorine is _____ . (1)

   d) All the elements are listed in the _____ . (1)

2. a) Give one example of a metal. (1)

   b) Give four characteristics of metals. (4)

3. a) Give one example of a non-metal. (1)

   b) Give four characteristics of non-metals. (4)

Compounds and molecules

Understand how a small number of elements can lead to millions of different compounds

- When **two or more elements** combine as a result of a **chemical reaction**, a compound is formed.
- When atoms of elements combine during a chemical reaction, the links between them are called **chemical bonds**.

Understand the idea of a molecule and simple formulae

- Particles in a compound are called **molecules**.
- These molecules are all the same in one particular compound, but they contain atoms of more than one element.
- Compounds are represented by formulae showing the atoms that make up each **molecule**.

Here are some examples:

H_2O is the formula for a **water molecule** which is made from two atoms of hydrogen and one atom of oxygen.

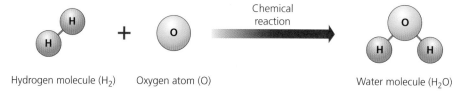

Hydrogen molecule (H_2)   Oxygen atom (O)   Water molecule (H_2O)

CO_2 is the formula for a **carbon dioxide molecule**, which is made from one atom of carbon and two atoms of oxygen.

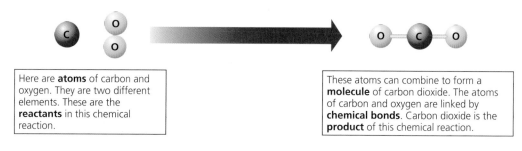

Here are **atoms** of carbon and oxygen. They are two different elements. These are the **reactants** in this chemical reaction.

These atoms can combine to form a **molecule** of carbon dioxide. The atoms of carbon and oxygen are linked by **chemical bonds**. Carbon dioxide is the **product** of this chemical reaction.

Know the formulae for these compounds

Formula	Compound
H_2O	water
CO_2	carbon dioxide
CH_4	methane
NaCl	sodium chloride (common salt)
HCl	hydrochloric acid
NaOH	sodium hydroxide
$CaCO_3$	calcium carbonate
$CuSO_4$	copper sulfate
H_2SO_4	sulfuric acid

Recognise that compounds have different properties to the elements from which they are made

- Compounds are **entirely different** in all respects from the elements that reacted to form them.
- The elements you start with are called **reactants**.
- The substance(s) produced as a result of a chemical reaction are called the **products**.
- The elements that are present as reactants will also be present in the products. They could be in a different combination or form, but they will be there.

For example:

hydrogen (H_2) + oxygen (O_2) $\rightarrow$ water (H_2O)

reactants   **product**

Recommended practical activities

1. **Making iron sulfide from its elements by heating iron and sulfur.**

   iron + sulfur $\rightarrow$ iron sulfide

   reactants   heat   product

2. **Compare the properties of iron (Fe), sulfur (S) and iron sulfide (FeS).**

Substance	Appearance	Action of a magnet	Conducts electricity?
Element: sulfur (S)	Yellow-green	None	No
Element: iron (Fe)	Grey-black	Magnetic	Yes
Compound: iron sulfide (FeS)	Black	None	No

■ Iron sulfide

Exam-style questions: Exercise 16

1 Use the following words to complete the sentences below.

chemical reaction   molecule   products   elements   compound

a) When two or more _____ combine as a result of a _____ , a _____ is formed. (3)

b) A water _____ is made from two atoms of hydrogen and one atom of oxygen. (1)

c) The substance(s) produced as a result of a chemical reaction are called the _____ . (1)

2 a) Copy the table below. Use what you know about elements and compounds to complete the table. (9)

Substance	Chemical symbol or formula	Solid, liquid or gas (at room temperature)	Colour	Does it conduct electricity?	Any special property?
oxygen					
carbon					
helium					
iron					
copper					
water					
iron sulfide					
methane					
sodium chloride					

b) Name the substances in the table that are compounds. (1)

- Remember that heating or cooling can cause a change of state. Four of these changes are: melting, boiling, condensing, evaporating (see Chapter 14: States of matter – the particle model.)
- Temperature is a measure of how hot or cold things are. Temperature differences can be felt or measured using a thermometer.
- The boiling point of pure water is 100°C and the freezing point is 0°C.
- The body temperature of a healthy human is 37°C.

Know that pure substances can be identified by their melting and boiling points

There are different ways you can identify whether a substance is pure or a mixture:

- **Visual** Sometimes it is easy to see that a substance contains different particles: for example, you can see that wet cement contains sand, gravel and water.
- **Physical properties** A magnet could be used to separate some metals from non-metals in a mixture.
- **Melting and boiling points** All substances have their own boiling and melting points. This fact can be used to:
  - identify a substance
  - determine whether it is pure or not.

A **pure** substance **always melts** at a certain temperature and **boils** at a certain temperature.

However, if impurities such as salt are added to pure water, the new water mixture will boil at a temperature higher than 100°C and freeze below 0°C. This is why salt is spread on roads when frost and ice is forecast.

Know that evaporation can happen at any temperature

Evaporation (liquid to gas) takes place at **any temperature**. This is because it depends on how much energy the particles have to break away from other particles and become a gas (vapour). (See Chapter 14: States of matter – the particle model)

Understand that pure substances comprise particles of the same type

- In a **pure substance** all the particles (atoms or molecules) are the same and, therefore, all samples of a pure substance behave in the same way.
- A **mixture** contains more than one type of particle. The particles are not joined together and so the mixture can behave in different ways. The behaviour will depend on the number and types of different particle in the mixture.

Know and understand the anomalous properties of water

We know the following about pure water:

- the boiling point is 100°C
- the freezing point is 0°C
- evaporation happens at any temperature.

When water is heated or cooled, it changes its state and the particles change position. It is unlike any other substance because:

- It **expands** when it freezes (causing burst water pipes in winter and freeze–thaw weathering in rocks).
- Frozen water (ice) floats on liquid water. This is because ice is less dense than water. The water molecules form a lattice structure in which they are more spread out than when they are in a liquid state.

Testing for pure water

Step 1: Boiling Test whether the liquid reaches its boiling point at 100°C. The boiling point for pure water is 100°C.

Step 2: Evaporation Check whether or not there is a **residue** after the liquid has evaporated. When pure water is evaporated to dryness, there is no residue.

Recommended practical activities

Recall how to:

- measure the melting point of a solid (such as stearic acid) and an impure substance (such as candle wax)
- investigate ice floating on water
- experiment with freezing plastic bottles full of water.

Exam-style questions: Exercise 17

1 Complete the following sentences.

  a) In a _____ substance all particles are the same and, therefore, all samples of the substance behave in the same way. (1)

  b) Pure water boils at _____ . (1)

  c) _____ of water happens at any temperature. (1)

  d) Adding impurities to water causes its boiling point to _____ . (1)

2 Name **two** unusual characteristics of water. (2)

3 Describe what happens to the particles in water as it changes state from liquid to water vapour. (1)

4 Amal is given a bottle of water labelled 'Pure London spring water'. She is curious to see if this is correct. Describe how she could investigate whether this claim is true or not. (2)

Mixtures including solutions

Know that the properties of a mixture are the same as those of its components

- A **mixture** contains more than one type of particle (see Chapter 17: Pure substances), which is put together without a chemical reaction having taken place. Making a mixture is a **physical** change.
- The particles are not joined together and so the mixture can behave in different ways.
- The behaviour will depend on the number of different particle types contained in the mixture.

Different particles have different properties. The properties of a mixture are the same as the properties of its components.

Here are some examples of mixtures:

Name of mixture	What is mixed
Any solution	solute (substance that dissolves) and solvent (substance solute dissolves in)
Any dilute acid	concentrated acid (solute) and water (solvent)
Blue ink	blue powder (solute) and water (solvent)
Seawater	salt and many chemicals (solute) and water (solvent)
Air	nitrogen, oxygen, carbon dioxide, water vapour

Understand that in a solution, particles are arranged randomly

- The different particles in a mixture are **not** joined together by chemical bonds.
- When a substance is dissolved in water it forms a solution. For example, a sugar cube dissolved in water forms a sugar solution.
- Remember these definitions:
  - **Solute** = a substance that dissolves
  - **Solvent** = the substance that a solute dissolves in
  - **Solution** = solute + solvent
- The formation of a solution is a **physical change**.
- In the example above, the sugar particles have not disappeared but are just randomly spread out amongst the water molecules. In fact, they are so spread out that we cannot see them. The solute and solvent particles are arranged randomly.
- These particles can be separated by physical methods (see Chapter 19: Separating mixtures).

Seawater, for example, is a solution containing randomly arranged particles of elements and compounds made up of elements such as oxygen, hydrogen, chlorine and sodium.

Know that air is a mixture of gases and know their proportions

Air is a mixture of elements and compounds.

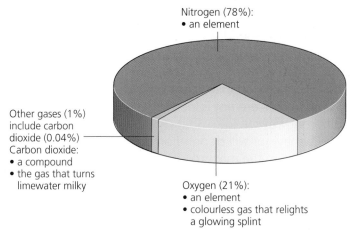

Nitrogen (78%):
• an element

Other gases (1%) include carbon dioxide (0.04%)
Carbon dioxide:
• a compound
• the gas that turns limewater milky

Oxygen (21%):
• an element
• colourless gas that relights a glowing splint

■ Proportions of different elements and compounds in air

Know the uses of oxygen and carbon dioxide in respiration and photosynthesis

The gases oxygen and carbon dioxide are essential to life.

Animals

Oxygen

- During breathing, the lungs take oxygen from the air and pass it into the blood.
- The blood takes oxygen to every cell in the body.
- The oxygen is used in **aerobic respiration**, a series of chemical reactions that release energy. Energy is needed for life processes to continue, otherwise death is inevitable.
- Respiration goes on in every living cell (see Chapter 8: Respiration).

Plants

Carbon dioxide

- Plants make their own food by the process of **photosynthesis** (see Chapter 7: Photosynthesis).
- They use carbon dioxide from the air and water from the soil.
- Carbon dioxide supplies carbon and oxygen – some of the ingredients needed for making carbohydrates.
- Carbohydrates are a store of energy and are used to make glucose, starch and cellulose, all of which are essential parts of plant cells.

Oxygen

- Some oxygen is used by the plant itself for **aerobic respiration** (see Chapter 8: Respiration).
- Oxygen not used by the plant will be released to the air.

Recommended practical activities

1. Mixture of iron and sulfur

- When you mix iron (Fe) and sulfur (S) no **chemical change** takes place.
- Iron (a metal) is magnetic and sulfur (a non-metal) is not magnetic.
- To separate these elements from the mixture, you can stir the powder with a magnet. The iron will stick to the magnet and the sulfur will not.

■ Separating iron from a mixture of sulfur and iron, using a magnet

2. Experiments to determine proportion of oxygen in air

Recall any investigations you may have done in class using, for example, a candle or rusting iron.

a) Using a candle

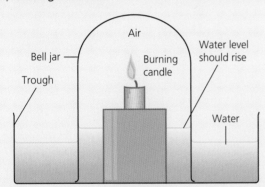

A candle needs oxygen to burn. In this experiment, the candle burns using up the oxygen in the air present in the bell jar. As the oxygen is used up and the volume of air in the bell jar decreases, the water level rises as the used oxygen is replaced by water. This change in volume equals the volume of oxygen in the air.

b) Rusting iron

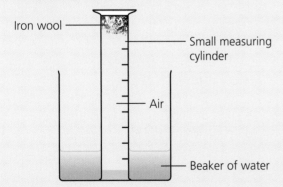

As the oxygen reacts with the iron, producing iron oxide (rust), oxygen is being taken from the air in the tube. This missing oxygen is replaced by water being pushed into the tube by air pressure. This can take around 3–4 days. The added volume of water in the tube equals the volume of oxygen in the air. You should find this is around 20%.

Exam-style questions: Exercise 18

1 Choose which option best completes the following sentences.

   a) Pure water is _____ .   (1)
      an element   a mixture   a compound   a solution

   b) An example of a mixture is _____ .   (1)
      pure water   sulfur   seawater   iron filings

   c) A substance that dissolves is called a _____ .   (1)
      solution   solvent   solute

   d) The substance a solute dissolves in is called a _____ .   (1)
      solution   solvent   solute

   e) The properties of a mixture are _____ the properties of its components. (1)
      different to   the same as

   f) Particles are arranged _____ in a solution.   (1)
      in a fixed pattern   randomly

2 Explain why oxygen is crucial to the life of animals.   (2)

3 Explain how the following gases are important to the lives of plants.
   a) oxygen   (1)

   b) carbon dioxide.   (1)

Dissolving

- Some solids (salt, sugar) **dissolve** in water to give solutions but some (sand, chalk) **do not**.
- Substances that dissolve in liquids are called soluble substances.
- Substances that do not dissolve in liquids are called insoluble substances.
- **Solubility** describes how well a substance dissolves: that is, how much solid dissolves in a particular amount of liquid (solvent).

When a substance dissolves, the particles of the substance being dissolved (the solute) are randomly distributed among the random arrangements of particles of the substance that is doing the dissolving (the solvent). You can see this when you dissolve hydrated copper sulfate crystals in water: the crystals disappear as a blue colour spreads throughout the clear solution.

- A solution that has the maximum amount of solute dissolved in it at a particular temperature is called a saturated solution. Remember that solubility of solids usually increases as the temperature rises.
- Undissolved solid on the bottom of a beaker will tell you that the solution is saturated.

Factors affecting the rate of dissolving

- Substances will dissolve more rapidly in warm water than in cold water.
- Small crystals or powders will dissolve more easily than large crystals.
- Stirring makes a substance dissolve faster.

Separating mixtures

Sieving

Sieving separates two or more solids with different-sized particles.

Filtering

- Insoluble solids can be separated from liquids by a process called filtration. Only solutions or pure liquids can pass through filter paper.
- **Residue** is the insoluble solid left in the filter paper – the mud left from filtering muddy water is the residue.
- **Filtrate** is the liquid that passes through the filter paper. It will be a solution or a pure liquid.

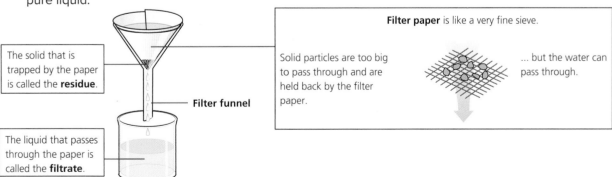

The solid that is trapped by the paper is called the **residue**.

Filter funnel

The liquid that passes through the paper is called the **filtrate**.

Filter paper is like a very fine sieve.

Solid particles are too big to pass through and are held back by the filter paper.

... but the water can pass through.

Decanting

Decanting is when a liquid is carefully poured off an insoluble solid. It is another way of separating insoluble solids from liquids.

Evaporation

Evaporation is the only way of removing the solvent from a solution.

As a solution is heated, the liquid (solvent) changes into a gas (evaporates), leaving the dissolved solids from the solution behind.

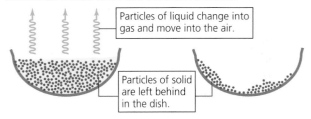

Particles of liquid change into gas and move into the air.

Particles of solid are left behind in the dish.

■ The process of evaporation separates solutes from solvent

Filtration and evaporation

A mixture containing a solvent, a solute and an insoluble substance can be separated into the different substances by using two methods in turn: filtration and evaporation.

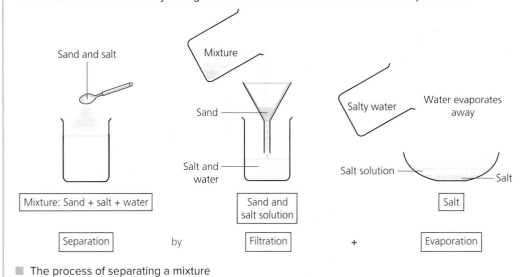

Sand and salt

Mixture

Sand

Salt and water

Salty water

Water evaporates away

Salt solution

Salt

Mixture: Sand + salt + water

Sand and salt solution

Salt

Separation   by   Filtration   +   Evaporation

■ The process of separating a mixture

Test your preliminary knowledge: Exercise 19A

1 You have a mixture of sulfur (a yellow solid) and hydrated copper sulfate (blue crystals). Sulfur is insoluble in water. Hydrated copper sulfate is soluble in water.

   a) What do the words soluble and insoluble mean? (2)

   b) What happens when you add water to your mixture? (2)

   c) If you filter the mixture after adding water, which substance is:

   i) the residue (1)

   ii) the filtrate? (1)

2 Copy and complete the following sentences.

   a) A sieve is used to separate two or more ____ with ____ -sized particles. (2)

   b) Filtering is used to separate ____ solids from ____ . (2)

   c) During filtration, the liquid that passes through the filter paper is called the ____ . (1)

   d) Gently pouring off clear liquid, leaving an insoluble solid behind is called ____ . (1)

3 Use words from the list below to complete the sentences.

   evaporated   solute   solvent

   a) When a solution is heated, the liquid (called the ____) changes into a gas. (1)

   b) When the liquid has ____ , the ____ is left behind. (2)

Know about distillation

Distillation is used to recover the solvent (the liquid) from a solution. This process can be used when the solvent is valuable and needs to be saved. The process has two parts:

● Evaporation – to remove the solvent from the solution as a vapour.
● Condensation – to change the solvent vapour into pure liquid solvent.

If one liquid is recovered, the process is called **simple distillation**.

Simple distillation

Simple distillation, to separate a solvent from a solution, can be undertaken using a water-cooled (Liebig) condenser. The condenser has two main features:

● The condensing tube is kept cool by an outer tube containing cold water moving in the opposite direction to that of the hot vapour flow.
● The condensing tube slopes down towards the collecting tube (or flask).

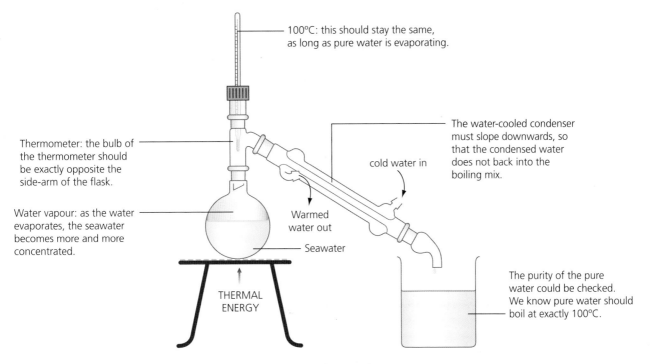

100°C: this should stay the same, as long as pure water is evaporating.

Thermometer: the bulb of the thermometer should be exactly opposite the side-arm of the flask.

Water vapour: as the water evaporates, the seawater becomes more and more concentrated.

The water-cooled condenser must slope downwards, so that the condensed water does not back into the boiling mix.

cold water in

Warmed water out

Seawater

THERMAL ENERGY

The purity of the pure water could be checked. We know pure water should boil at exactly 100°C.

Simple distillation – producing pure water from seawater

Understand that substances with different boiling points are separated by distillation

Distillation works because the substances involved have different boiling points.

The dissolved solute has a higher boiling point than the solvent.

In the example above, as the seawater is heated, water vapour evaporates from the solution.

Recommended practical activities

Recall:

- the distillation of pure water from seawater (see previous page)
- the effect of evaporating tap water, pure water and seawater (see previous page).

Exam-style questions: Exercise 19B

1 Make a table with the headings 'elements', 'compounds' and 'mixtures'.
   Put the following substances into the correct columns: (12)

   air   carbon   carbon dioxide   distilled water
   seawater   iron filings   crude oil   dilute sulfuric acid
   magnesium   oxygen   sodium chloride   iron sulfide

2 a) Draw a labelled diagram to show how you would best obtain some
       pure water from a sample of seawater. (6)

   b) How would you show that the water you had obtained was pure? (2)

3 What method would you use to:

   a) recover solid salt from salt solution? (1)

   b) recover mud from muddy water? (1)

Know the technique of chromatography and about using different solvents

Chromatography is a method used to separate mixtures of two or more soluble substances. The method depends upon the physical property of solubility.

- Some solids dissolve better than others; every substance has a different solubility that may change with temperature.

Water is not the only **solvent** you will have used, although it is the solvent used in most writing inks. You might have used **ethanol** or **propanone** to extract the green colour from leaves and dry cleaners will use special solvents to remove grease stains from clothes.

- The solvent will move up chromatography paper (which is like filter paper) and will take the dissolved solids with it.
- The chromatography paper tries to absorb the solids and 'hold them back'.
- Each dissolved solid will move at different speeds (and slower) on the paper, while the solvent continues to travel upwards.

Understand the factors which affect the separation of spots and how it can be used for identification

In the experiment below, the separating solvent moves up through the paper, dissolving the different substances from the spots.

- The **most soluble** substance travels the **furthest** up the paper.
- The **least soluble** substance travels the **shortest distance**.

In the example below, an unknown sample X is being tested against spots of known pure substances, so it can be identified. We can see from the results that sample X is a mixture of pure substances C and D.

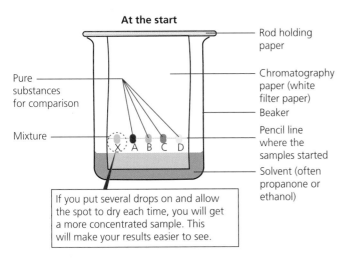

At the start

Pure substances for comparison

Mixture

If you put several drops on and allow the spot to dry each time, you will get a more concentrated sample. This will make your results easier to see.

X A B C D

Rod holding paper

Chromatography paper (white filter paper)

Beaker

Pencil line where the samples started

Solvent (often propanone or ethanol)

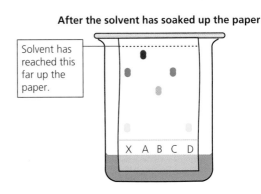

After the solvent has soaked up the paper

Solvent has reached this far up the paper.

X A B C D

■ Using chromatography to identify unknown substances in a mixture

- Chromatography is used for analysis of small quantities of mixtures. It will often be used by:
  - food chemists – to check the additives and ingredients in food (for example, the colourings in sweets) are correctly listed
  - police – to match up samples from a suspect to substances found at a crime scene
  - biochemists – to separate proteins into their various parts in the constant search for new drugs to cure illnesses
  - lateral flow tests used to diagnose an infection, for example, COVID-19, use the chromatography technique. The liquid sample, taken from the patient, is dropped into the base of the test. It then travels up the paper. If the infection is present, as it meets the testing chemical in the paper, two red lines appear.

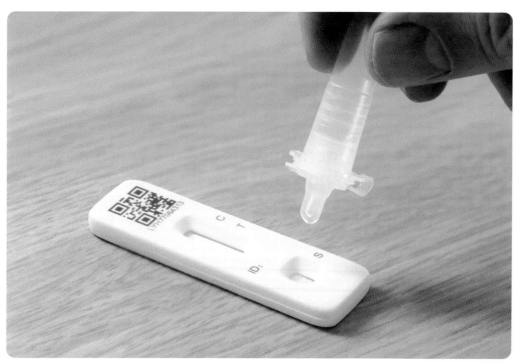

■ Lateral flow test

Recommended practical activities

Recall any investigations where you have used the chromatography technique to separate and identify mixtures of different soluble substances in, for example, felt-tip pens or coloured sweets, using either water or propanone as the solvent.

Exam-style questions: Exercise 20

1 Describe the method you would use to find out how many pigments there are in the ink from a water-based black felt-tipped pen. (4)

2 In a chromatography experiment, substance A travels further up a filter paper than substance B. Explain why this happens. (1)

3 Which of the following could be investigated using paper chromatography?
**compounds in drugs   pollutants in water   sand and water mixture
blood samples** (1)

Preliminary knowledge

Remember:

- Virtually all materials, including those in living systems, are made through chemical reactions.
- Chemical reactions happen all around us, for example, ripening fruit, setting superglue, cooking food.
- In most cases, chemical reactions are permanent and cannot be reversed.
- A chemical reaction means a new substance has been made.
- Materials are made as a result of both:
  - naturally occurring processes (for example, oxygen and sugars in plants are made by photosynthesis)
  - human-made processes (for example, plastic, concrete, medicines and laundry liquid are manufactured, synthetic, substances).
- If new properties are really useful, then humans may be able to deliberately change how some materials behave.

Know how to use a Bunsen burner

A Bunsen burner is the most common source of heat in a laboratory. Look at the diagram below and remind yourself of how to use a Bunsen burner safely.

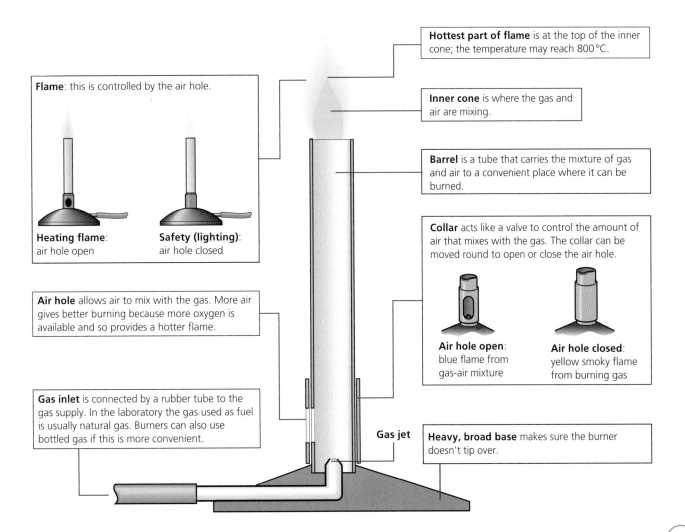

Hottest part of flame is at the top of the inner cone; the temperature may reach 800 °C.

Inner cone is where the gas and air are mixing.

Barrel is a tube that carries the mixture of gas and air to a convenient place where it can be burned.

Collar acts like a valve to control the amount of air that mixes with the gas. The collar can be moved round to open or close the air hole.

Air hole open: blue flame from gas-air mixture

Air hole closed: yellow smoky flame from burning gas

Flame: this is controlled by the air hole.

Heating flame: air hole open

Safety (lighting): air hole closed

Air hole allows air to mix with the gas. More air gives better burning because more oxygen is available and so provides a hotter flame.

Gas inlet is connected by a rubber tube to the gas supply. In the laboratory the gas used as fuel is usually natural gas. Burners can also use bottled gas if this is more convenient.

Gas jet

Heavy, broad base makes sure the burner doesn't tip over.

- Substances **burn** when they are heated in air. Burning is a chemical change called **combustion**.
- Combustion is a chemical reaction in which energy in the chemical store of energy in the fuel (reactant – see below) is transferred to a thermal store of energy in the surroundings and a chemical store of energy in the resulting compound (products – see below).
- Combustion needs heat to get it started. Here are some examples:

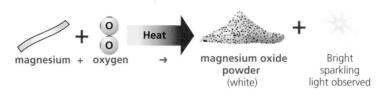

magnesium + oxygen   →   magnesium oxide powder (white)   Bright sparkling light observed

■ Combustion of magnesium (Mg)

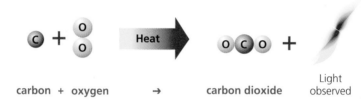

carbon + oxygen   →   carbon dioxide   Light observed

■ Combustion of carbon (C)

Recognise the need for oxygen

- Oxygen is a reactive gas: responsible for burning, **rusting** and respiration in living things.
- During combustion, oxygen from the air combines with another element to form an oxide. This combining reaction is called **oxidation**.
- Fuels (reactants) by themselves do not transfer energy. It is the combustion reaction with oxygen that transfers the energy.

Understand the idea of a reaction as a rearrangement of particles – conservation of mass

There are two types of reaction that can take place:

Physical changes

These include changes of state (see Chapter 14: States of matter – the particle model) and dissolving. During a physical change:

- substances do not change into other substances; there is no chemical reaction
- particles remain unchanged but they do change positions relative to each other
- no chemical bonds are formed and no particles are lost or gained
- no change in chemical formulae.

Physical changes are temporary and may be reversed: melted butter will become solid when it cools; salt may be recovered from a solution by evaporation (see Chapter 19: Separating mixtures).

The law of conservation of mass

● No particles are lost or gained in a physical change, so the total mass of the substances involved does not change.

Chemical reactions

In a chemical reaction, a substance or substances change into a new substance or substances.

reactants → products

For example, heating copper carbonate:

copper carbonate → copper oxide + carbon dioxide

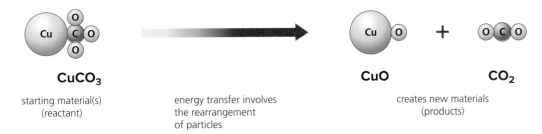

$CuCO_3$		CuO	CO_2
starting material(s) (reactant)	energy transfer involves the rearrangement of particles	creates new materials (products)	

● In most cases, the change is permanent and cannot be reversed.
● Light may be emitted while energy is transferred from one store to another.
● Thermal energy (from a chemical store) is often transferred to the environment.
● The molecules are rearranged.
● The formulae change.

The law of conservation of mass

Even when atoms interact and create new products in a chemical reaction, mass is conserved. Atoms join together in different ways to form products. The new substances created are made up of the atoms that were present in the reactants. No new atoms have been created and none have been destroyed. Mass is conserved.

Know the tests for oxygen, carbon dioxide and water

● When oxygen reacts with an element, the element is oxidised: only one product is formed. In the oxidation of magnesium (see earlier in the chapter) only magnesium oxide is formed.
● When oxygen reacts with compounds, more than one product is produced.

For example, when a candle (made from a compound, of carbon and hydrogen) is burned it produces water and carbon dioxide:

candle + oxygen → water + carbon dioxide

Both the hydrogen in the water and the carbon in the carbon dioxide must have come from the candle.

We can test for the presence of these substances using the following tests:

Test for oxygen

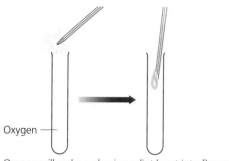

Oxygen will make a glowing splint burst into flame.

Test for carbon dioxide using limewater

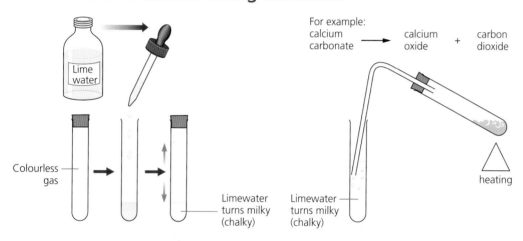

For example:
calcium carbonate → calcium oxide + carbon dioxide

Colourless gas

Limewater turns milky (chalky)

Limewater turns milky (chalky)

heating

Test for water using anhydrous copper sulfate

Anhydrous copper sulfate is WHITE.   Copper sulfate is BLUE.

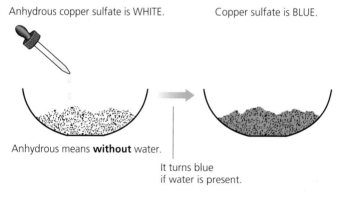

Anhydrous means **without** water.

It turns blue if water is present.

Understand and devise word equations

Chemical reactions can be described by word equations with an arrow (→) used to mean 'change into'.

The word equations for the combustion reactions we saw at the start of this section can be written as:

- Burning of magnesium in oxygen

  magnesium + oxygen → magnesium oxide

- Burning of carbon in oxygen

  carbon + oxygen → carbon dioxide

Recommended practical activities

Recall any investigations you have done to explore:

- the burning of elements in air and oxygen
- mass changes when burning magnesium.

Exam-style questions: Exercise 21

1 a) Explain what is meant by **combustion**. (1)

   b) What **two** substances does combustion require? (2)

2 What does the word **oxidation** mean? (2)

3 Explain the law of conservation of mass. (1)

4 Write down the word equation for the burning of sulfur in oxygen. (1)

5 Explain why a CO_2 fire extinguisher will extinguish an electrical fire. (1)

Ask yourself

The incidence of wildfires has increased over the last few years. Think about what impact this has on the Earth's ecosystems and atmosphere.

22 Fuels and production of carbon dioxide

Know that hydrocarbon compounds such as coal are used as energy sources

- Burning fuels (in the presence of oxygen) is an important chemical change.
- Fuels include:
  - living fuels – wood and materials from plants
  - semi-fossil fuels – peat
  - fossil fuels – coal, oil, natural gas. These are fuels formed from the dead remains of animals and plants that lived millions of years ago.
- Energy is transferred in the burning process.
- The thermal store of energy can warm our homes, heat our water and cook our food.
- Light given out during burning is used to help us see in the dark.

Hydrocarbons

Compounds that contain only hydrogen and carbon are called **hydrocarbons**.

hydrocarbon + oxygen → water + carbon dioxide

- Light might be emitted while energy is transferred from one store to another.
- Energy (from a chemical store) is often transferred to the environment.
- Some hydrocarbons come from crude **oil**.
- They are found in the many products made from crude oil, including petrol, candle wax, polythene and some plastics.
- Nearly all forms of transport burn some form of hydrocarbon as a fuel.
- Natural **gas** is the hydrocarbon methane. If your laboratory is connected to the gas supply, this is the gas that supplies your Bunsen burners.

Coal

- Fossil fuels include coal, oil and gas.
- Power stations are able to transfer the chemical store of energy in coal as electricity.

Understand how the products of combustion are formed

Most fuels contain carbon compounds. This means that when they burn, carbon dioxide is released into the atmosphere. The exception to this is hydrogen, which produces water vapour when it burns.

Burning coal

- Coal contains compounds of both carbon and sulfur.
- When coal is burned, both carbon dioxide and sulfur dioxide (another gas) are released into the atmosphere.
- Sulfur dioxide dissolves in the water of the atmosphere to form acid rain.

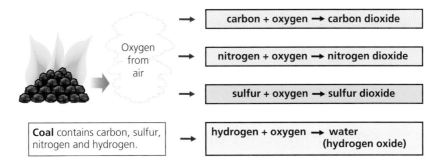

Oxygen from air

carbon + oxygen → carbon dioxide

nitrogen + oxygen → nitrogen dioxide

sulfur + oxygen → sulfur dioxide

hydrogen + oxygen → water (hydrogen oxide)

Coal contains carbon, sulfur, nitrogen and hydrogen.

Burning gas

Natural gas is hydrocarbon methane. It burns creating two products – water and carbon dioxide.

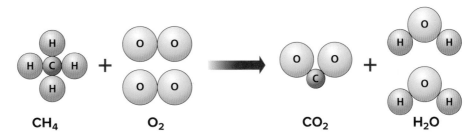

CH_4 O_2 CO_2 H_2O

Burning oil

- Many useful hydrocarbons come from crude oil that has been separated into its various compounds by fractional distillation at a refinery.
- Incomplete burning of petrol causes carbon monoxide (a poisonous gas) to be released.

Recognise the impact on climate of burning fossil fuels – climate change

Although burning fossil fuels transfers a lot of energy, it causes problems too:

- Fossil fuels are non-renewable.
- Fossil fuels pollute the atmosphere. The products of combustion, smoke, ash and other waste gases harm the quality of air. Carbon dioxide and several other of these products are major factors in **climate change**.
- Earth is warmed by radiation from the Sun. When the radiation reaches Earth, it is reflected back into the atmosphere. However, 'greenhouse' gases (such as carbon dioxide and methane) stop some of this reflected heat from escaping and it is absorbed.
- Burning fossil fuels produces more and more greenhouse gases, so more heat is absorbed by the atmosphere, increasing its temperature (global warming) and leading to climate change.

Likely effects of global warming

- Increasingly strong winds.
- More deserts.
- More flooding.
- Melting ice caps.
- Heavier rain.
- Pests can spread to new warmer areas.
- Forest fires.

Climate change is now a major concern to much of the global community. Action needs to be taken to reduce the amount of greenhouse gases produced.

Actions needed to combat global warming

- Reduce carbon emissions worldwide.
- Reduce the burning of fossil fuels.
- Find alternative, non-polluting, energy resources.
- Reduce the cutting down of forests.
- Plant more forests.

Know that sulfur dioxide and carbon monoxide are polluting gases and understand how they are formed

- Coal contains the elements carbon, sulfur, nitrogen and hydrogen.
- When coal is burned, these elements combine with oxygen in the air to form the oxides carbon dioxide, nitrogen dioxide (mainly formed at high temperatures) and sulfur dioxide.
- Water is also a product of this combustion.
- When the air pollutants nitrogen dioxide and sulfur dioxide react with water and air, they form airborne sulfuric and nitric acids.

Sulfur dioxide

sulfur + oxygen → sulfur dioxide

Sulfur dioxide is a gas and is a main component of acid rain.

Carbon monoxide

Carbon monoxide, a gas, forms from the incomplete combustion of fuels such as coal, oil, petroleum, natural gas or wood. It can cause serious health problems, or death if it is inhaled.

Recognise the causes and effects of acid rain

- When sulfur dioxide combines with water in clouds it forms sulfuric acid.
- Nitrogen oxide, another product of combustion of some fuels, also combines with water to form nitric acid.
- When it rains, this acid falls to the ground and is called acid rain.
- Acid rain can cause many problems:
  - Damage to environments, or death such as lakes and rivers, which affects fish and other organisms in these habitats.
  - Damage to forests: acid rain damages the soil as well as affecting the trees directly.
  - Damage to human-made structures, particularly those made of limestone or marble.

Recommended practical activity

1. Test the products of a burning candle.

When a candle burns in air, there are two products: carbon dioxide and water.

candle wax + oxygen → water + carbon dioxide

Candle wax contains only hydrogen and carbon and is, therefore, a hydrocarbon.

Exam-style questions: Exercise 22

1 Methane is a hydrocarbon.

   a) What is a hydrocarbon? (1)

   b) Which two products are formed when methane burns? (2)

   c) Write a word equation to describe the burning of methane. (4)

2 Coal is a fossil fuel that burns well in air.

   a) What is meant by the term **fossil fuel**? (1)

   b) Name **two** other fossil fuels. (2)

   c) i) Water vapour is one gas released when coal burns in air. Name **two** others. (2)

   ii) Suggest what happens to each of these gases when they come into contact with water vapour in the air. (2)

3 Describe how the combustion of fuels provides a source of energy. (2)

4 a) Explain how the burning of fossil fuels contributes to global warming. (4)

   b) Give four of the likely effects of global warming on the climate. (4)

   c) What two actions could be taken to address this issue? (2)

   d) Explain how clearing forests affects carbon dioxide levels. (2)

Ask yourself

If you were in charge, what steps would you take to reduce the amount of acid rain that falls on Earth?

As we have seen in previous sections, new substances can be formed during chemical reactions. The new substances formed when two or more elements combine are called **compounds**.

In this section, we are looking at how metals react with certain substances.

Know the reaction of metals with oxygen

Reactions with oxygen are called **oxidation reactions**.

When metals react with oxygen, they form **metal oxides**.

Most of these reactions need heat to get them started.

metal + oxygen → metal oxide

Metal	Result of burning	Metal oxide formed (product)
magnesium	Burns with a brilliant white flame	magnesium oxide (grey-white powder)
zinc	Burns with a bright, fierce blue-green flame	zinc oxide (yellow-grey)
iron	Burns with yellow sparks	iron oxide (black powder)
copper	Very hard to burn – small green flame is produced	copper oxide (black powder)
gold	Does not burn at all	

Know the reaction of metals with water

The word equation to show how metals react with water:

metal	+	water	→	metal hydroxide	+	hydrogen
(solid)		(liquid)		(liquid)		(gas)

Metals react to different degrees.

- Some metals react very vigorously when placed in water.
- Less reactive metals need steam to form the oxide and hydrogen.
- Some metals (including copper and gold) do not react with water at all.

Metal hydroxides are alkalis (bases).

Metal	Reaction with water	Hydroxide or oxide formed (product)
magnesium	These reactions are very slow in cold water but can be sped up using **steam**. Hydrogen is given off slowly	magnesium hydroxide
zinc		zinc oxide
iron		iron oxide
copper	No reaction	

Know the reaction of metals with acid

All acids are compounds of hydrogen.

metal + acid → salt + hydrogen

For example:

magnesium + sulfuric acid → magnesium sulfate + hydrogen

When conducting experiments, always use the same acid. You could use sulfuric acid or hydrochloric acid.

Metal	Action of cold dilute acid	Salt formed with dilute hydrochloric acid
magnesium	Reacts to form salt and hydrogen	magnesium chloride
zinc	Reacts slowly to form a salt and hydrogen	zinc chloride
iron	Reacts slowly to form a salt	iron chloride
copper	No reaction	

Know the test for hydrogen – a squeaky pop!

For example:

hydrochloric acid + zinc → zinc chloride + hydrogen

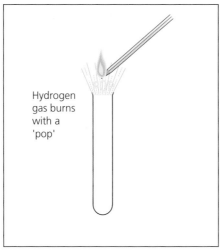

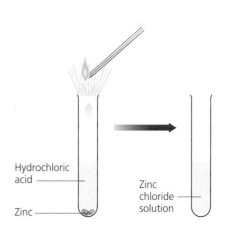

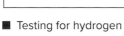
■ Testing for hydrogen

Know about rusting of iron and understand how it can be prevented

Rusting occurs when iron is exposed to oxygen and water. You may remember putting some wet iron wool in a measuring cylinder and inverting this over water. After a time, the iron has rusted and the water has risen up the inside of the cylinder. When the reaction stops – after a few days – you will see that 20% of the air has been used up (in the same way as 20% of the air is used up when burning a candle in a bell jar), showing the involvement of oxygen in rusting.

Rusting costs money

- To replace items that have rusted.
- To take measures to prevent rusting from taking place.

There are several ways of preventing iron from rusting

- Exclude oxygen and water by covering with a layer of grease, oil, paint or plastic.
- Exclude oxygen and water by covering the iron with a metal, such as tin or zinc.

Recommended practical activities

1. Reactions of magnesium

Magnesium (Mg) is a very reactive metal. Recall how it reacts with air, water and acid:

- Magnesium reaction with air – burns easily

  magnesium + oxygen → magnesium oxide

- Magnesium reaction with cold water – reacts slowly

- Magnesium reaction with steam – reacts quite vigorously

  magnesium + water + magnesium hydroxide + hydrogen

- Magnesium reaction with acid – reacts

  magnesium + sulfuric acid → magnesium sulfate + hydrogen

2. Rusting

Investigate the necessity for both air and water for the rusting of iron.

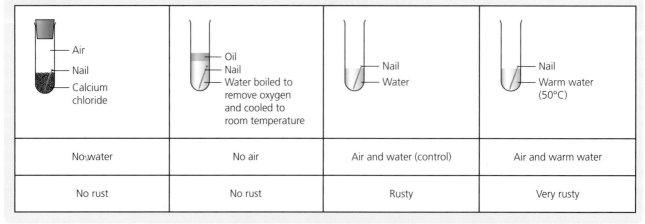

Air Nail Calcium chloride	Oil Nail Water boiled to remove oxygen and cooled to room temperature	Nail Water	Nail Warm water (50°C)
No water	No air	Air and water (control)	Air and warm water
No rust	No rust	Rusty	Very rusty

Exam-style questions: Exercise 23

1 Explain what is meant by oxidation. (1)

2 Look at the following and say whether a reaction will take place. If it does, name the products.

   a) Magnesium heated in air. (1)

   b) Magnesium in water. (1)

   c) Copper placed in dilute hydrochloric acid. (1)

   d) Zinc placed in sulfuric acid. (1)

3 Why is copper a good metal to use when making saucepans? (1)

4 a) Why is rusting said to be one of the most expensive chemical reactions? (2)

   b) Name the three substances involved in rusting. (3)

   c) Describe three ways to help prevent rusting taking place. (3)

Remember:
- A mixture contains more than one type of particle, which is put together without a chemical reaction taking place.
- Mixtures can usually be separated quite easily (see Chapter 19: Separating mixtures and Chapter 20: Combustion – chemical reactions). This is because a mixture is created by a **physical change**.
- A chemical reaction that can break down a compound is called a decomposition reaction.
- When the energy needed to break down a compound is provided by **heat**, the process is called **thermal decomposition**.

Know that heating brings about decomposition and recognise the products formed

Copper carbonate

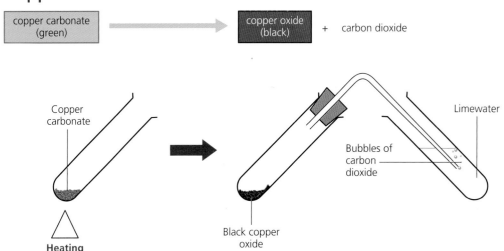

The thermal decomposition of copper carbonate forms copper oxide and carbon dioxide

Calcium carbonate

Calcium carbonate (found in limestone) needs to be heated very strongly for the thermal decomposition reaction to happen.

calcium carbonate → calcium oxide + carbon dioxide

Hydrated copper(II) sulfate

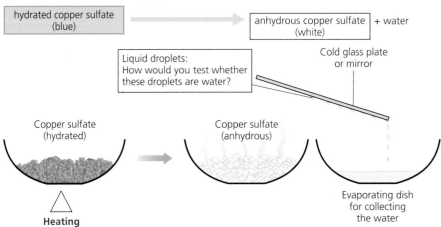

The thermal decomposition of hydrated copper(II) sulfate forms anhydrous copper(II) sulfate and water and carbon dioxide

Know that some substances do not change chemically when heated

Some compounds cannot be decomposed by heating. Copper oxide (CuO) is an example of a compound for which heating does not cause a chemical change.

Recommended practical activity

Recall what happens when copper carbonate and hydrated copper sulfate are heated (see previous page).

Exam-style questions: Exercise 24

1 Complete the following sentences.

   a) A chemical reaction that can break down a compound is called
     a _____ reaction.   (1)

   b) When heat is needed to break down a compound, the reaction
     is called a _____ reaction.   (1)

2 Copper carbonate is heated in an experiment. What test would you use to identify the gas being released?   (1)

3 In the thermal decomposition reaction of hydrated copper(II) sulfate, droplets of liquid are produced. How would you test whether these droplets are pure water?   (2)

> **Preliminary knowledge**
>
> - Solutions can be classified as acidic, neutral or alkaline.
> - Indicators can be used to classify solutions as acidic, neutral or alkaline. They indicate that a particular substance is present.
> - Indicators change their colour when they come into contact with particular substances.

Know the pH scale

Testing 'strength' of acids and alkalis – the pH scale

The **pH scale** runs from 0 to 14.

pH scale (a scale of numbers ranging from 1 to 14).

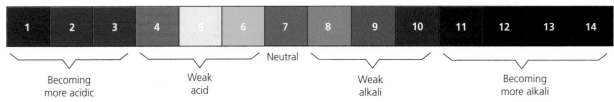

The pH scale showing colours using Universal Indicator (a scale of numbers from 1–14)

- The middle of the scale is 7 and all liquids with a pH of 7 are **neutral**.
- Solutions that have a pH less than 7 are **acids**.
- Solutions that have a pH more than 7 are **alkalis**.
- Weak acids/alkalis have pH values close to the mid-point (7). For example, a range of 5–9 includes:
  - 'natural' acids found in foods including vinegar and citrus fruits (pH 5–7)
  - 'weak' alkalis such as detergents, medicine to cure indigestion, baking powder (pH 7–9)
- Strong acids/alkalis have pH values close to the ends of the scales (0 and 14). For example:
  - sulfuric and hydrochloric acids (pH 0–4)
  - sodium and calcium hydroxides (pH 10–14)
- Alkalis are the chemical opposite of acids.

Know the indicators

Plant dyes

Red plants such as red cabbage and beetroot can be boiled in water to produce a purple solution that will act as an indicator.

Litmus tests

Litmus is extracted from a type of lichen and forms a purple solution in water.

- Purple litmus solution will turn red when it is added to acids.
- Purple litmus solution will turn blue when it is added to alkalis.

Litmus is used in the laboratory in three different forms:

- As a purple solution of the dye in water.
- As red test papers to detect alkalis.
- As blue test papers to detect acids.

Universal Indicator

- pH values can be determined by using Universal Indicator.
- This is a mixture of plant dyes and is available as a green liquid or as test papers.
- A few drops of Universal Indicator are added to a substance.
- The indicator shows a different colour for different pH values.

Know about the use of a pH meter

A pH meter is an electronic device used to measure the pH values of solutions. It is particularly useful when an exact pH value is needed. Universal Indicator cannot do this because the colours could be seen differently by different people or in different light.

Recognise neutralisation reactions

Acids are able to 'cancel out' alkalis in a chemical reaction known as neutralisation (see Chapter 26: Neutralisation reactions). When equal quantities of acid and alkali of the same strength and concentration are mixed, a neutral solution is formed.

For example:

- Hydrochloric acid (hydrogen chloride in water) is a strong acid – Universal Indicator added to it goes red.
- Sodium hydroxide is a strong alkali – Universal Indicator added to it is purple.
- Add the acid slowly to the alkali until the neutral point is reached – Universal Indicator turns green.
- The resulting solution is a neutral solution of sodium chloride (a salt), dissolved in water.
- This can be shown by the word equation:

acid	+	alkalis	→	salt	+	water
hydrochloric acid	+	sodium hydroxide	→	sodium chloride	+	water

- The water can be evaporated from the sodium chloride solution to leave crystals of solid sodium chloride.

Know the acid/base properties of metal oxides and non-metal oxides

Remember chemicals can be put into two groups:

Group **1**   Acids – pH less than 7
Group **2**   Alkalis and bases – pH more than 7

A **base** is a substance that can neutralise acids. The following are all bases:

- metal oxides
- carbonates
- hydroxides

An alkali is a base that can dissolve in water. The following are all alkalis:

- ammonium hydroxide
- sodium hydroxide
- calcium hydroxide (limewater)

Recommended practical activities

Recall investigations where you have:

- tested acids and alkalis with litmus and Universal Indicator
- measured pH with Universal Indicator and a pH meter
- added alkali to acid while measuring pH.

Exam-style questions: Exercise 25

1 What colour will purple litmus paper go when put into the following liquids? (5)

   a) hydrochloric acid

   b) limewater

   c) sodium hydroxide

   d) sugar solution

   e) water

2 Draw a pH scale running from 0 to 14. On your diagram label:

   a) the neutral point (1)

   b) acids – showing strongest and weakest (2)

   c) alkalis – showing strongest and weakest. (2)

3 a) What is a base? (1)

   b) Give two examples of a base. (2)

   c) What is an alkali? (1)

   d) Why is toothpaste alkali? (1)

4 a) Write a word equation to show a neutralisation reaction. (1)

   b) Give one example of a neutralisation reaction. (1)

   c) Wasp stings are alkaline. What could you use to alleviate the pain? (1)

   d) Bee stings are acidic. What could you use to alleviate the pain? (1)

Acids are able to 'cancel out' alkalis in a chemical reaction known as **neutralisation**.

Neutralisations produce a **salt** and water.

Know reactions of acids with metals

Some metals react with acids. When they do, salts are formed and the gas given off is hydrogen.

> **metal + acid → salt + hydrogen**

For example:

> **zinc + sulfuric acid → zinc sulfate + hydrogen**

Know reactions of acids with alkalis

The word equation to describe the chemical reaction when an acid reacts with an alkali:

> **acid + alkali → salt + water**

Acids are neutralised by bases and alkalis.

Know reactions of acids with metal oxides

All metal oxides are called bases.

When a base neutralises an acid, the name of the base becomes the first name of the salt.

> **base + acid → salt + water**

> **metal oxide + acid → salt + water**

For example:

> **copper oxide + sulfuric acid → copper sulfate + water**

> **iron oxide + sulfuric acid → iron sulfate + water**

> **magnesium oxide + hydrochloric acid → magnesium chloride + water**

> **zinc oxide + hydrochloric acid → zinc chloride + water**

Know reactions of acids with metal carbonates

Metal carbonates can make salts. Metal carbonates produce **carbon dioxide** when added to acids and they neutralise acids to form salts.

> **metal carbonate + acid → salt + water + carbon dioxide**

For example:

> **copper carbonate + sulfuric acid → copper sulfate + water + carbon dioxide**

Recognise the formation of salts

Neutralisation produces a salt and water.

The word equation to describe the chemical reaction that has happened is:

acid + alkali → salt + water

Recognise the effect of acidity on the environment

- Rainwater is naturally acidic due to carbon dioxide dissolved in the water.
- Rainwater (acid rain) contains weak acids sulfuric and nitric acid. (See Chapter 22: Fuels and production of carbon dioxide to revise the effects of acidity on the environment.)

Know about the uses of limestone

Buildings

Limestone (a base) contains the compound calcium carbonate. It can be used for:

- construction of buildings
- production of cement.

Agricultural lime

Agricultural lime is crushed limestone that is used to treat acid soils. Over time, soils have become more acidic which has a negative impact on crop productivity. The addition of lime to the soil works by dissolving and releasing a base that lowers the acidity of the soil. If the soil is more alkali (basic) plants will be able to absorb more nutrients from the soil. It also supplies calcium, magnesium and other minerals to crops.

Other uses

- Limestone is a major ingredient in toothpaste.
- It can be used as a food additive to provide calcium for strong teeth and bones.

Know about the weathering effect of acid rain on limestone

Carbon dioxide dissolves in rainwater to make a weak acid (carbonic acid). In polluted areas, rainwater may also contain small amounts of other acids including dilute sulfuric acid and nitric acid. If acid rain falls on limestone, which contains calcium carbonate, a chemical reaction takes place.

calcium carbonate + sulfuric acid $\longrightarrow$ **calcium sulfate + water + carbon dioxide**

Buildings or objects, such as headstones, made from limestone become weathered because the rain slowly dissolves them. Acid rain speeds up this process. Rain can destroy limestone buildings and cause caves to appear in areas of limestone rock. This is known as limestone weathering.

Recommended practical activities

Investigate the reaction of an acid such as hydrochloric acid with alkalis, metals, metal oxides and metal carbonates.

Exam-style questions: Exercise 26

1 Copy and complete the following:

   a) acid + base → _____ + water (1)

   b) acid + alkali → _____ + water (1)

   c) acid + carbonate → _____ + water + _____ (2)

   d) acid + metal → _____ + _____ (2)

   e) Adding acid to an alkali _____ the pH of the mixture. (1)

   f) Neutralising an acid by a base _____ the pH of the mixture. (1)

   g) Adding water to an acid _____ the pH of the acid. (1)

2 Describe each of the following reactions by:

   i) writing the general word equation for each reaction

   ii) providing an example of each reaction.

   a) The reaction of hydrochloric acid with an alkali. (4)

   b) The reaction of hydrochloric acid with a metal. (4)

   c) The reaction of hydrochloric acid with a metal oxide. (4)

   d) The reaction of hydrochloric acid with a carbonate. (4)

3 Explain how you would test for the gas carbon dioxide. (2)

4 Explain how you would test for the gas hydrogen. (2)

Ask yourself

If limestone is damaged by acid rain, why do you think it is chosen as a building stone?

■ Limestone erosion

Chemistry – Test yourself

Before moving on to the next chapter, make sure you can answer the following questions. The answers are at the back of the book.

1 **a)** Using what you know about particles, complete the following table:

	Solids	Liquids	Gases
Arrangement of particles			
Do they flow easily?			
Can they be compressed?			
Can they change their shape?			

    b) Explain the changes of state below in terms of what happens to the particles.
       i) Ice melting.
      ii) Water boiling.
    c) Explain the term sublimation.
    d) **i)** Explain gas pressure in terms of the movement of particles.
      ii) Suggest what might happen to the air pressure inside a football when it is left outside in the sun.

2 **a)** Complete the following sentences:
      i) Atoms can join together to form _____ .
      ii) A substance made of one type of atom is called an _____ .
     iii) A chemical symbol represents _____ atom of that element.
     iv) In some elements, the particles are molecules: two or more _____ are bonded together.
     v) The Periodic Table lists 100 elements, which fall into two categories: _____ and _____ .
    b) Describe how you test different elements for their conductivity.

3 **a)** Complete the following sentences:
      i) When two or more elements combine as a result of a chemical reaction a _____ is formed.
      ii) The links between the atoms of elements are called _____ .
    b) **i)** Give an example of a compound and write down its formula.
      ii) For the compound you have chosen in **i)**, give the elements from which it is made.
     iii) In a chemical reaction, do the reactants have the same properties as the product?

4 **a)** Complete the following sentences:
      i) A _____ contains more than one type of particle without a chemical reaction taking place.
      ii) Making a mixture is a _____ change.
     iii) Particles in a mixture are not joined by _____ bonds.
     iv) Properties of a mixture are _____ the properties of its components.
    b) Suggest how you would separate a mixture of iron and sulfur.

5 **a)** Explain when you would use distillation to separate substances.
    b) When is this a good technique to use?
    c) Give an example of a mixture you could separate in this way. Draw a diagram to explain the method you would use.

6 a) Complete the following sentences:

    i) Burning is a chemical change called ____ .

    ii) Combustion needs ____ to get it started.

    iii) During combustion, ____ from the air combines with another element to form an ____. This reaction is called ____ .

b) Give an example of a combustion reaction by writing down a word equation.

7 Coal contains the elements carbon, sulfur, nitrogen and hydrogen.

    a) Explain how sulfur dioxide is formed and write out the word equation.

    b) i) Give the name of the substance formed when the product from **a)** combines with water

       ii) When it rains, this substance falls from the sky. What is it commonly known as?

    c) Give two examples of how acid rain can damage the environment.

8 a) i) Write a word equation to show how metals react with water.

       ii) Write the word equation showing what happens to magnesium when it reacts with water.

    b) i) Write a word equation to show how metals react with acid.

       ii) Give an example of this type of reaction.

9 When energy, provided in the form of heat, is needed to break down a compound, the process that occurs is called thermal decomposition.

    a) Write the word equation to show the thermal decomposition reaction of copper carbonate.

    b) Draw a diagram to explain the apparatus and method you would use to conduct this experiment.

    c) How would you test if the gas you collect is carbon dioxide?

10 a) What does an indicator do?

    b) Name **two** indicators and say what they are used for.

    c) i) What colour does red litmus paper turn when dipped in hydrochloric acid?

       ii) What colour does red litmus paper turn when dipped in calcium hydroxide?

    d) A base is a substance that can neutralise acids.

       i) Write out a word equation to describe an example of this chemical reaction.

       ii) Give an example of a neutralisation reaction. Your answer must include the correct chemical names for each substance.

    e) Some metals react with acids.

       i) Write out a word equation to show an example of this reaction.

       ii) Give an example.

    f) Metal oxides (bases) react with acids.

       i) Write out a word equation to show an example of this reaction.

       ii) Give an example.

    g) Metal carbonates (bases) react with acids.

       i) Write out a word equation to show an example of this reaction.

       ii) Give an example.

Know that the Sun is the ultimate source of energy

Most of the Earth's energy comes from the Sun.

The role of plants

- Plants use energy transferred by light from the nuclear store of energy in the Sun to the chemical stores of energy in plants (see Chapter 7: Photosynthesis).
- Only green plants can make their own food by photosynthesis; they are called **producers**.
- The store of energy in plants enables them to grow and increase in biomass.
- Energy is transferred between chemical stores of energy in one organism in the chain to another (see Chapter 11: The interdependence of organisms in an ecosystem).

Plants provide a source of fuel in several ways:

- **Fossil fuels** are produced when plants and animals die and become trapped in certain conditions. Over millions of years, they can become coal, oil and natural gas (see below).
- **Wood** from trees is an important fuel for burning.
- **Alcohol** is produced from fermenting plants.

Solar panels

- Make direct use of the Sun's energy.
- When light from the Sun hits the solar panels, it interacts with semiconductors to produce an electric current.
- Collecting energy from the Sun does not change the amount of energy that can be collected from the Sun in the future, so this is a renewable energy resource (see below).

Know about the range of fossil fuels

Fossil fuels are a chemical store of energy. Energy is transferred to other energy stores, most often including a **thermal store of energy**, when fuels are burned.

- **Coal** is used mainly in power stations to generate electricity.
- **Natural gas** is mainly methane and is used in homes and in power stations.
- **Butane** and **propane** are stored as liquid for use in camping stoves.
- **Oil** is the source of useful substances including:
  - **kerosene** for domestic heating and aircraft fuel
  - **petrol** and **diesel** for vehicles
  - many molecules needed for the production of synthetic materials, including most plastics.

Understand the difference between renewable and non-renewable energy resources and their advantages and disadvantages

Renewable energy sources

Renewable energy resources **do not get smaller** as they are used.

- **Solar power:** Sunlight hits solar panels and interacts with semiconductors to produce an electric current or is used to heat water.
- **Biomass:** Animal and plant waste can be used to generate biogas (mostly methane).

Whatever energy is available must be transferred into an energy store that can be used. Electricity transfers the kinetic store of energy from a turbine turning a generator to other energy stores. A turbine may be turned using:

- **Wind power:** Wind (kinetic store of energy) to turn the fan of a turbine.
- **Geothermal power:** Electric power generated using steam or very hot water. It is only available in areas with hot springs and geysers.
- **Hydroelectric power:** Water trapped behind a dam holds energy as a gravitational (potential) store. The water is released to drive turbines. It is only available in mountainous areas.
- **Tidal power:** Water trapped at high tide also holds energy as a gravitational (potential) store that can be transferred to a kinetic store of energy in turbines.
- **Wave power:** The up and down movement of waves is used to drive a turbine.

Advantages of renewable energy resources	Disadvantages of renewable energy resources
They are not used up	Can be unreliable: for example, output falls if there is no wind, not much sun or not many waves
Little or no atmospheric pollution	Only available in certain places
They are becoming cheaper to provide	Building and maintaining solar panels and turbines can be expensive
	Can be bad for the environment: for example, wind and solar farms take up significant amounts of land; wind turbines can affect migrating birds; damming rivers for hydroelectric schemes can affect fish populations and destroy habitats in valleys

Non-renewable energy sources

Fossil fuels

Advantages of non-renewable energy resources	Disadvantages of non-renewable energy resources
• Often easy to extract (oil and gas) • Can be used at any time	• Supplies are running out • Take millions of years to form, so they cannot be replaced • Very polluting

Nuclear power

Energy is transferred from a nuclear store of energy by splitting uranium atoms in nuclear reactors. It is currently classified as a sustainable energy resource because it uses mined uranium, the source of which is finite. The use of this nuclear store of energy does not generate greenhouse gases, although other pollutants, such as radioactive waste, are produced. The advantages and disadvantages include:

- zero-emissions, so good for air quality
- small land footprint compared to the area needed by wind and solar farms
- produces less waste than a coal-fired power plant but the waste has to be carefully handled and processed.

It is argued that nuclear energy could provide large amounts of low-carbon power alongside renewables to help reach net-zero targets and provide the energy the world needs.

■ Dungeness Nuclear Power station, UK

Understand that a non-renewable resource is one that cannot be replenished within a lifetime

Non-renewable resources have taken millions of years to form so cannot be replaced within a lifetime.

As plants and animals die, they decompose. Over millions of years they are covered by layers of sediment and get buried deeper and deeper within the Earth's crust. The heat and pressure turn them into fossil fuels such as oil and gas.

■ A large quarry showing the vertical layers of coal

Recognise that a variety of processes are used to generate electricity

Energy stored in fuels can be transferred into other stores of energy. As we have seen, electricity, the energy pathway by which energy is transferred from one store to another, can be generated in a number of ways:

- by burning (fossil fuels and biomass)
- by semiconductors in solar panels (solar power)
- by turbines and generators (wind, hydroelectric, tidal, geothermal power)
- by the splitting of atoms (nuclear).

Recommended practical activities

Recall any of the following practical activities you may have done in class:

- Observe solar cells in operation to light a light-emitting diode (LED) or power a motor or buzzer.
- Investigate the uses of fossil fuels and possible alternatives.
- Research the use of renewable energy resources to generate electricity or provide heating.

■ Solar panel street lighting

■ Solar panel water pump used for irrigation

■ Wind turbines generating electricity

■ Ground source heat pump

Exam-style questions: Exercise 27

1 a) What does the term fossil fuel mean? (2)

b) Name two fossil fuels. (2)

c) Why are fossil fuels called non-renewable? (2)

d) Is burning the only use of fossil fuels? (2)

e) Explain the advantages and disadvantages of fossil fuels. (4)

2 a) What does the term renewable energy mean? (2)

b) Name two renewable energy resources. (2)

c) Explain the advantages and disadvantages of renewable energy resources. (4)

Energy transfers and conservation of energy

Know that energy resources can be measured and stored in a number of different ways

Energy is the measure of the ability to do:

- work that has to be done
- work that is able to be done.

The standard unit of energy is the **joule** (J). Large amounts of energy are measured in **kilojoules** (kJ).

1 kJ = 1000 J

Energy resources or energy stores can exist in different forms.

Store of energy	Energy stored in:	Examples
Gravitational (potential) store	an object at height	aeroplane, a person on top of a ladder
Kinetic store	moving objects	moving car, kicked football, running rhino
Chemical store	chemical bonds	food, batteries, muscles, petrol
Internal (heat/thermal) store	hot objects where particles vibrate faster	hot cup of coffee
Elastic store	an object that is squashed or stretched	compressed spring, stretched elastic band
Nuclear store	the nucleus of an atom	the Sun, uranium

■ The Sun

Understand that energy is conserved

Energy can be transferred from one store to another store. **In every energy chain, the total amount of energy stays the same** even though it is being transferred from one store to another.

Law of conservation of energy:

total amount of energy stored at the start = total amount of energy stored at the end

Understand that when energy is transferred from one store to another, it is sometimes via a pathway or process such as electricity, light or sound

Energy is transferred from one store to another via energy pathways. Here are some examples:

Example	from one store	via pathway	to another store
Burning coal	Chemical store of energy in fuel	Light*	Thermal store of energy
Shaking a baby's rattle	Chemical store of energy in baby's body	Sound	Kinetic store of energy in surroundings
A clock striking	Kinetic store of energy in the striker transferred to a kinetic store of energy in the vibrating bell	Sound to our ears	Thermal store of energy in the surroundings
Electric heater	Chemical store of energy in fuel (in power station)	Electricity	Thermal store of energy in the surroundings
Torch being switched on	Chemical store of energy in the battery	Light	Thermal store of energy in the surroundings

Energy is transferred by chemical reaction from the chemical store of the coal and the oxygen in the air to the thermal store of the surroundings and the chemical store of the carbon dioxide that is made. Some of the thermal store of energy is transferred by light to the thermal store of more distant surroundings.

Recognise that, although energy is always conserved, it may be dissipated, reducing its availability as a resource

Although energy is always conserved, some energy is transferred or dissipated to the surroundings as a thermal store of energy. The thermal store of energy in the surroundings increases.

For example: a television screen works using electricity to transfer energy to the electromagnetic and thermal energy stores within in the screen. The light emitted allows patterns on the screen to be seen. Since a television is not used to heat the room, you could say that the energy being transferred to the thermal store of energy is no longer available for use by the television: its availability as a resource has been reduced.

Energy store in fuel transferred via electricity (input)

Useful device

Useful energy store (output)

This can be used to carry out **work** that is useful to humans.

This energy store is **dissipated.**

Wasted energy: transfer to thermal store (heat) and sound

Recommended practical activities

1. **Observe and carry out different energy conversions, for example:**
   - heating substances
   - using solar energy to light a LED
   - dropping a ball and describing the energy changes.

2. **Research where and how energy is wasted in power stations and cars.**

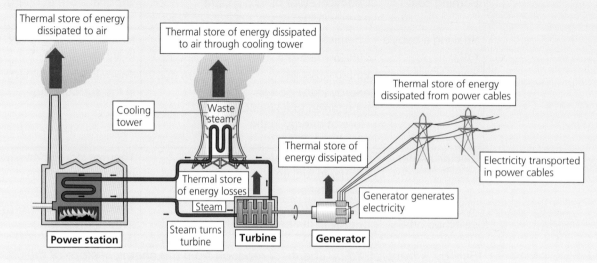

■ Diagram to show how energy is dissipated (transferred) as waste in a power station

Exam-style questions: Exercise 28

1 Match the type of energy store with where it might be found. (6)

Energy store	Found in
A Gravitational store of energy	**1** Nuclei of atoms
B Elastic store of energy	**2** A food mixer
C Nuclear store of energy	**3** Food
D Chemical store of energy	**4** A catapult
E Kinetic store of energy	**5** A hammer hitting a nail
F Thermal store of energy	**6** An electric heater

2 Explain the law of conservation of energy. (2)

3 Describe the energy transfers that take place when:

   a) you strike a match (2)

   b) a skier skis down a mountain (2)

   c) an electric light is turned on. (2)

29 Speed and movement

Know that speed is measured using a variety of units (often m/s) but that it is always the distance covered in a given time

Speed is a measure of how fast something is moving.

We need to measure:

- the distance the object moves in metres (m) or kilometres (km)
- the time it takes to move this distance in seconds (s) or hours (h).

Speed is the distance moved in each time period, so it is measured in metres per second (m/s) or kilometres per hour (km/h).

Know the quantitative relationship between speed, distance and time

You can calculate speed using the following equation:

$$\text{speed} = \frac{\text{distance}}{\text{time taken}} \qquad s = \frac{d}{t}$$

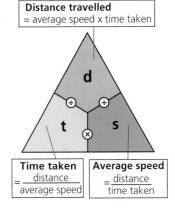

Distance travelled
= average speed x time taken

Time taken
= $\frac{\text{distance}}{\text{average speed}}$

Average speed
= $\frac{\text{distance}}{\text{time taken}}$

Example 1: Calculate speed

Car A travelled a distance of 100 km in 2 hours. If it travelled at a constant speed, what was its speed?

$d = 100\,\text{km} \qquad t = 2\,\text{h}$

$$s = \frac{d}{t}$$

$$= \frac{100\,\text{km}}{2\,\text{h}}$$

$$= 50\,\text{km/h}$$

Car A is travelling at a speed of 50 km/h.

Example 2: Calculate distance

Use the same formula in the form: **distance = speed × time**

Cecelia walked at a speed of 3 km/h for 5 hours. What distance did she cover?

$s = 3\,\text{km/h} \qquad t = 5\,\text{h}$

$d = s \times t$

$\quad = 3\,\text{km/h} \times 5\,\text{h}$

$\quad = 15\,\text{km}$

Cecelia covered a distance of 15 km.

Example 3: Calculate time

$$time = \frac{distance}{speed}$$

Kabir cycled a distance of 40 km at a speed of 10 km/h. How long did this take?

$d = 40\,km \qquad s = 10\,km/h$

$$t = \frac{d}{s}$$

$$= \frac{40\,km}{10\,km/h}$$

$$= 4\,h$$

It took Kabir 4 hours to cycle 40 km.

Note: be very careful to use the correct units.

Understand how to measure and calculate the speed of a moving object

To measure speed you need to:

- measure the exact distance travelled (in the correct units)
- measure the length of time taken (in the correct units).

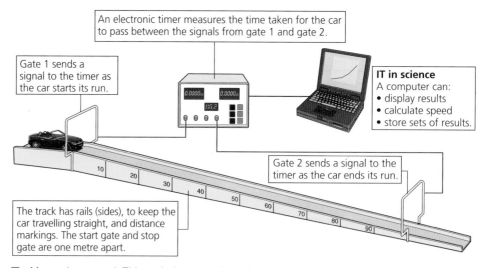

An electronic timer measures the time taken for the car to pass between the signals from gate 1 and gate 2.

Gate 1 sends a signal to the timer as the car starts its run.

IT in science
A computer can:
- display results
- calculate speed
- store sets of results.

Gate 2 sends a signal to the timer as the car ends its run.

The track has rails (sides), to keep the car travelling straight, and distance markings. The start gate and stop gate are one metre apart.

■ Measuring speed. This technique can be adapted to measure the speed of other things.

Recognise that speeds for vehicles approaching or passing each other in a straight line add or subtract as seen from one vehicle

- When two cars are travelling side by side, it may appear the faster one moves past slowly even though both vehicles are travelling at a high speed. This is called **relative motion**, and is how objects move relative to each other.
- You can calculate the relative speeds of two objects as follows:
    - Objects moving in the **same direction**

    relative speed = fastest speed – slowest speed

    - Objects moving **towards each other**

    relative speed = speed object A + speed object B

Recommended practical activities

- Recall any measurements you have done to calculate the speeds of different vehicles, for example:
  - a trolley or model car in the lab
  - a falling ball
  - runners on a track
  - cars passing the school gates.

Exam-style questions: Exercise 29

1 In running races, athletes are timed electronically.

   a) What is the formula you would use to calculate speed? (1)

   b) Calculate the mean average speed for the runners in the table below. Give your answers to one decimal place. (4)

Race distance, in m	Time taken, in s	Speed, in m/s
100	9.58	
800	101	
5 000	755	
10 000	1 571	

2 a) How far would you travel in 30 s if you ran at 7 m/s? (1)

   b) How long will a walk take if you travel 500 m at 2 m/s? (1)

3 Two cars are travelling in the same direction on a motorway. Car A is travelling at 65 km/h and is overtaken by car B, which is travelling at 86 km/h. Calculate their relative speed. (1)

4 You are investigating the mean speed at which different objects fall after being dropped from a height of 4 metres.

   a) Describe what equipment you would need. (4)

   b) Describe the method you would use. (4)

   c) Show how you would calculate the mean speed from the information you have collected. (2)

   d) Explain how you would make sure it is fair test. (1)

30 Measuring force

What is a force?

- A **force** is either a **push** or **pull**.
- A force can:
  - make stationary things move
  - make a moving object go faster or slower (**change speed**)
  - make a moving object **change direction**
  - make a moving object stop
  - change the **shape** of an object.

Key facts about forces

- All forces:
  - have size
  - act in one direction only.
- We measure the size of a force in **newtons** (N) often using a newton meter (newton spring balance).

Types of force

Magnetic forces

(See Chapter 38: Magnetic fields and electromagnets.)

Gravitational force

- **Gravity** is the force of attraction between any masses.
- Gravity only pulls.
- The size of gravitational force depends upon:
  - the masses of the object
  - the distance between the centres of the objects.
- Gravity pulls all objects towards the centre of the Earth, so all objects experience a downwards force which is called the object's **weight**.

Push and pull forces

- A push is a force moving an object away from you.
- A pull is a force moving something towards you.

Frictional (including air resistance) force

Contact

- **Friction** occurs when any two surfaces rub together. Even the smoothest surfaces have ridges and high spots when looked at through a microscope. The high spots will try to stick together and the ridges oppose movement.
- Friction is the force that **opposes motion**.
- Friction may be reduced by lubrication: putting a layer of fluid between surfaces to keep the rough surfaces apart. For example:
  - oil – bicycle chain
  - water – ice skates
  - Air is also often used to reduce friction: for example, in a hovercraft.

Air resistance

Anything moving through the air will rub against air particles that will slow down the movement in a special type of friction called **air resistance**.

Support (reaction force)

When an object rests on a surface, the surface pushes up against the object. This **reaction force** from the surface balances the force of the object pushing downwards.

For example: If you are sitting on a chair, the gravitational force pulls you downwards towards the centre of the Earth. The upwards reaction force stopping you sinking downwards is sometimes called the **support force**.

Upthrust

Another reaction force is **upthrust**. Upthrust is the upward support force created by a liquid or gas. An object will float in water if the downward gravitational force pulling the object towards the centre of the Earth is balanced by the upthrust caused by the water. For example:

- A boat floats because the gravitational force pulling the boat downwards is balanced by the upward support force (upthrust) exerted by the water on the boat.
- A boat will sink if the gravitational force pulling the boat downwards is greater than the upthrust pushing it up.

Test your preliminary knowledge: Exercise 30A

1 List the three main effects of a force. (3)

2 Complete the sentences using the words below.

   newtons   friction   push   gravity

   a) A _____ is a force moving an object away. (1)
   b) The size of a force is measured in _____ . (1)
   c) _____ is the force of attraction between any masses. (1)
   d) _____ occurs when any two surfaces rub together. (1)

3 Explain why a boat floats on water. (2)

Know the unit used to measure force is the newton (N)

The size of a force is measured in **newtons (N)**.

A newton meter or force meter is an instrument used to measure force.

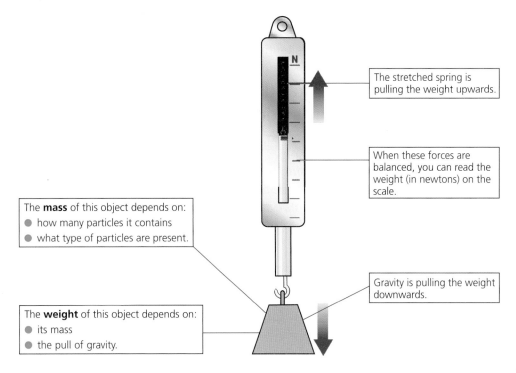

The stretched spring is pulling the weight upwards.

When these forces are balanced, you can read the weight (in newtons) on the scale.

The **mass** of this object depends on:
- how many particles it contains
- what type of particles are present.

Gravity is pulling the weight downwards.

The **weight** of this object depends on:
- its mass
- the pull of gravity.

Know that weight is a force caused by gravity and that weight = mass × gravitational field strength (g)

- Gravity pulls all objects towards the centre of the Earth. So all objects will experience a downwards force, which is called the object's **weight**.
- Weight is measured in newtons (N) using a newton meter.
- The force with which a 1 kg mass is pulled towards the centre of the Earth, or any other large mass, is called the **gravitational field strength** (g).
- On Earth, gravity pulls each kg of mass with a force of about 10 N so we say $g \approx 10$ N/kg ($\approx$ means 'approximately equal to').
- This is confirmed by the use of a newton spring balance (or newton meter).

Mass, in kg	Gravitational field strength, in N/kg	Weight (reading on newton meter), in N
1	10	10
2	10	20
0.5	10	5

From these results, we can see that:

weight (N) = mass (kg) × gravitational field strength (N/kg)

Understand that forces arise from the interaction between two objects

- Forces arise when two objects interact.
- Forces between objects can be caused by contact (for example, when you push a car) or because of a field, such as a gravitational field or a magnetic field.
- If an object is not moving then the forces must be balanced.
- **Gravity** creates a force of attraction (a pull) between two objects.
  - The greater the **mass** of the objects, the greater the gravitational attraction will be.
  - The size of the gravitational force also depends on the **distance** between the objects. The further they are away from each other, the smaller the force of gravity between them.

Understand that forces can cause stretching or compression

- Forces can cause elastic materials/objects to stretch or be compressed (squashed).
- **Stretching** happens when the material/object is pulled.
- **Compression** happens when the material/object is squashed.
- When this happens the space between the particles is changed.

For example:

- When weights are applied to a spring or rubber band, it stretches. The force being applied can be measured in newtons.

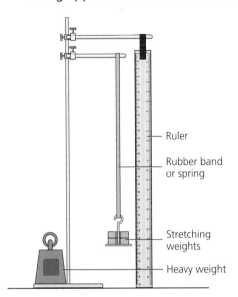

— Ruler

— Rubber band or spring

— Stretching weights

— Heavy weight

■ Experiment to investigate how the extension (stretching) of a material is affected by the amount of force that is applied

- When a catapult band is pulled back ready to fire, it stretches.
- When a tennis ball hits the ground, it gets compressed and then returns to its original shape.

Recognise that friction and air resistance are examples of an unbalanced force

- Forces act in pairs in opposite directions.
- When the forces acting on one object are **equal**, the forces are **balanced** – no change in movement happens.
- If **one force is bigger** than the other, the forces are **unbalanced** – movement happens in the same direction as the larger force acts.

NO MOVEMENT

MOVEMENT TO THE RIGHT

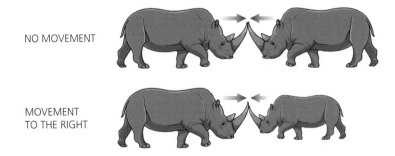

Friction

- When two surfaces rub together:
  - the force of friction opposes the motion
  - energy is transferred to a thermal store of energy.

For example, people camping might rub two sticks together to generate enough heat to light a fire.

Air resistance

Air resistance (drag) is a type of friction that occurs between an object and molecules in the air.

For example:

● When a parachutist leaves an aeroplane: weight > air resistance.
● The parachutist speeds up as he/she falls because the forces are unbalanced.
● When the parachute opens, air resistance increases and eventually becomes equal to the weight (see Chapter 31: Effects of forces).
● The parachutist falls at a steady speed: the forces are balanced so there is no change in movement.

Consider the effect of both of these forces in a lorry travelling along a straight road. It will move at a constant speed as long as the force provide by the engine balances the air resistance and friction from the road.

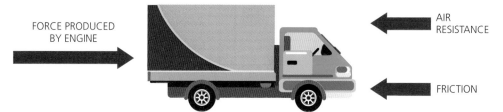

FORCE PRODUCED BY ENGINE

AIR RESISTANCE

FRICTION

Items such as cars, aeroplanes and cycle helmets are designed to reduce air resistance and unwanted friction. They are specially shaped (streamlined): their bodies have pointed fronts, sleek designs and smooth surfaces.

Recommended practical activities

Recall how you would measure the stretching or compression of a spring when weights are added (see the experiment set up on page 121).

Exam-style questions: Exercise 30B

1 On Earth, gravity exerts a force of 10 N/kg. Calculate the weight of the following masses:

   a) 3 kg (2)

   b) 30 kg (2)

   c) 180 kg (2)

   d) 500 g (2)

   e) 320 g (2)

2 On Earth, gravity exerts a force of 10 N/kg; on the Moon gravity exerts a force of 1.6 N/kg. An astronaut has a mass of 80 kg on Earth.

   a) What is his mass on the Moon? (1)

   b) What is his weight on Earth? (2)

   c) What is his weight on the Moon? (2)

Ask yourself

Look back at your laboratory notebook at the experiments you will have done rolling cars down slopes to see how far they ran over different surfaces. What did they tell you about which surface offered the most friction?

- Frictional forces, including air resistance, affect motion (think about, for example, streamlining cars, friction between tyres and road).
- The distance you need to bring a car to a stop is called the **stopping distance**. It has two parts:
  - The distance travelled between when you first think about applying the brakes to when you actually apply the brakes is called the **thinking distance**.
  - The distance travelled between when the brakes are applied and the car coming to a rest is called the **braking distance**.
- The stopping distance increases if a car is travelling faster because:
  - The car will travel further in the time it takes you to think about using the brakes (see Chapter 29: Speed and movement) – the thinking distance increases.
  - The faster a car is travelling, the more time the frictional force will take to stop the car – the braking distance also increases.
- If the road is wet, there is less friction between the tyres and the road so the car needs an even longer distance to come to a stop.
- In the UK, the Highway Code gives drivers guidelines on these distances.

Know that an unbalanced force will cause a change of speed or direction of an object and a balanced force will cause no change

- When a force acts, the other force of the pair acting in the opposite direction is called the **reaction force**.
- If an unbalanced force acts on a moving object in the opposite direction from the movement, then the speed of the object will decrease; in other words, the object will slow down and may eventually stop.
- If an unbalanced force acts on a moving object in the same direction as the movement, then the speed of the object will increase – it will go faster.
- When the two opposing forces are equal, they are balanced and the object will move at a constant speed in the same direction or remain at rest.

Here are some examples:
- Floating and sinking
  - Water exerts an upward force (upthrust) on all objects.
  - Gravity exerts a downward force (weight) on all objects.
  - When these two forces are balanced (equal), the object floats.
  - If weight is greater than upthrust, the object sinks.

- The force of gravity increases the speed of falling objects (unless air resistance = weight).
- Pedalling harder makes a bicycle go faster.
- If the downward force of gravity on a cat sitting still on a branch is balanced by the support force of the branch, the cat stays where it is.

Understand how to represent balanced and unbalanced forces by the use of force arrows in one dimension

- We show the direction and size of a force by drawing arrows:
  - longer arrow = big force
  - short arrow = small force.
- When forces work on the same object, the bigger force (longer arrow) wins.

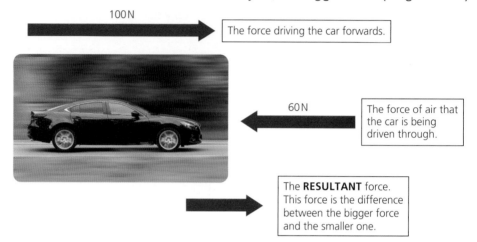

100N

The force driving the car forwards.

60N

The force of air that the car is being driven through.

The **RESULTANT** force. This force is the difference between the bigger force and the smaller one.

Recognise that friction and air resistance have a number of advantages and disadvantages

Friction can have advantages

- Brake pads are designed to produce as large a frictional force as possible, to slow down or stop a moving vehicle.
- Rough road surfaces and tyres with patterned ridges (tread):
  - enable cars and cycles to move
  - make them easier to control.
- Carpets are safer to walk on than shiny, polished floors.
- We can make our hands warmer by rubbing them together.

Friction can have disadvantages

- Friction can cause damage to objects such as brake pads on cars and the chain of a bicycle. The friction between the surface of the chain and the surface of the cogs on a bicycle, for example, causes layers of molecules to be rubbed off, resulting in wear. Using oil to lubricate the chain can help reduce friction and therefore lessen the damage.

Air resistance can have advantages

- As a parachutist opens the parachute, air resistance increases. The resultant force is now upwards and the parachutist slows down. This decreases the air resistance until it balances the weight. The parachutist travels at a steady speed that is slow enough for her to survive when she hits the ground.

Air resistance can have disadvantages

- It can slow aeroplanes down.
- It can slow vehicles down. The engines need to produce a greater force to allow them to overcome the air resistance and keep moving or speed up.
- Air resistance on a bird's body can slow it down, so every now and then it needs to go into a dive to maintain its speed.

Recommended practical activities

- Investigate the effect of an unbalanced force on a trolley or model car.
- Investigate a model parachute with different canopy areas.
- Research the thinking, braking and stopping distances given in the Highway Code.

Exam-style questions: Exercise 31

1 Write down the word or words that best complete each of the following sentences.

   a) If an unbalanced force acts on a moving object in the same direction as the movement, then the speed of the object will _____ (the object will _____). (2)

   b) If an unbalanced force acts on a moving object in the opposite direction from the movement, then the speed of the object will _____ . (1)

   c) When the two opposing forces are equal (that is, they are balanced), the object will move at a _____ or _____. (2)

2 Draw a lorry accelerating away from a junction.
  Draw the force arrows to show the forces acting on the lorry. (2)

3 In deep space, there is no friction and no gravity.

   a) Give the name of a force that causes a spaceship to speed up as it approaches Earth. (1)

   b) When the spaceship prepares to land on water, a parachute is deployed.

      i) Name the friction force that increases when the parachute opens. (1)

      ii) The spaceship lands on water. Describe the balanced forces that result in the spaceship floating. (2)

4 Describe how you would investigate how canopy area affects the speed of a model parachute. (6)

Ask yourself

Why do you think the body shapes of Formula 1 racing cars are designed as they are?

Pressure is the force per unit area. It depends on two things:

- how big the force is
- how large an area the force is working on.

The force acting on each square metre (m^2) or square centimetre (cm^2) is known as pressure.

Know the quantitative relationship between force, area and pressure

To calculate the pressure exerted by one surface on another, we use the relationship:

$$\text{pressure} = \frac{\text{force}}{\text{area}} \qquad P = \frac{F}{A}$$

If we apply the same amount of force:

- concentrating the force over a **larger area** gives a **lower pressure**
- concentrating the force over a **smaller area** gives a **higher pressure**.

This formula can be rearranged. If you need to calculate:

- **area** when you know the pressure and force, it becomes

$$\text{area} = \frac{\text{force}}{\text{pressure}}$$

- **force** when you know the pressure and area, it becomes

force = pressure × area

Know that the unit of pressure is N/m² or N/cm²

Force is measured in **newtons** (N).

Area is measured in **square metres** (m^2) or **square centimetres** (cm^2).

Therefore, using the relationship above, the unit of pressure is **newtons per square metre** (N/m^2) or **newtons per square centimetre** (N/cm^2).

Understand how to use the relationship for simple quantitative work

Here are some examples of how you can calculate pressure.

Force = 100 N

Area = 4 m²

$$\text{pressure} = \frac{100 \text{ N}}{4 \text{ m}^2}$$

$$= 25 \text{ N/m}^2$$

■ This cube exerts a pressure of 25 N/m²

Force = 100 N

Area = 2 m²

$$\text{pressure} = \frac{100 \text{ N}}{2 \text{ m}^2}$$

$$= 50 \text{ N/m}^2$$

■ In this example, the same 100 N force is applied to an area half the size of that in the first example, so the pressure is now 50 N/m²

Note: make sure you always give your answers in the correct units.

Recognise applications of pressure, for example, skis, snowboards, sharp objects

Concentrating the force on a **small area increases the pressure**. This can be useful and is applied in situations such as:

- Studs or spikes on sports shoes prevent wearers slipping on the field or track.
- A knife blade or sharp point concentrates the force from your hand over a small area, so cutting or piercing is made easy.

Spreading the force over a **large area decreases the pressure**. Things that make use of this include the following:

- Caterpillar tracks on tanks and bulldozers allow heavy machinery to travel over muddy ground without sinking.
- Skis and snowshoes spread the wearer's weight over a large area so they do not sink into soft snow.

Recommended practical activities

Recall any practical activities you may have undertaken to investigate pressure. These might include:

- measuring and calculating the pressure exerted by different-shaped objects
- investigating the effects of pressure in situations such as shoes on snow/sand and heels on floors
- researching the way in which tyre pressure affects the area of contact between road and tyre.

Exam-style questions: Exercise 32

1 a) What is the formula for calculating pressure? (1)

   b) What formula would you use to calculate force if you were given values for area and pressure? (1)

   c) State the unit of pressure. (1)

2 It is autumn and a farmer is ploughing a field to make it ready for planting the next season's crop. The plough cuts furrows in the earth creating holes for the seeds to be planted.

   a) Explain why the tractor has very large, wide and chunky tyres. (1)

   b) The blades on the plough are thin and sharp. Explain why they are designed like this. (1)

   c) The blade of the plough has an area of 2 cm² and exerts a force of 1200 N. Calculate the pressure the blade will exert on the soil.
      Show your working. (2)

33 Density

Know the relationship between density, mass and volume and how to use this for simple quantitative work

Density is how much mass is packed into each unit of volume of material. For example, a kilogram of feathers will take up more space than a kilogram of gold.

- Solids tend to be more dense than liquids.
- Liquids tend to be more dense than gases.

The equation for calculating density is:

$$\text{density} = \frac{\text{mass}}{\text{volume}} = \frac{m}{v}$$

Density (D) is in grams per cubic centimetre (g/cm³) or kilograms per cubic metre (kg/m³).

Mass (m) is in grams (g) or kilograms (kg).

Volume (v) is in cubic centimetres (cm³) or cubic metres (m³).

Simple calculations

- Calculating density of an object when you know the volume and mass:

This cube represents a sugar cube with a mass of 3 g.

(i) Calculate the density of one sugar cube.

mass = 3 g

volume = 2 × 2 × 2 = 8 cm³

$$\text{density} = \frac{\text{mass}}{\text{volume}}$$

$$= \frac{3}{8}$$

$$= 0.375 \, \text{g/cm}^3$$

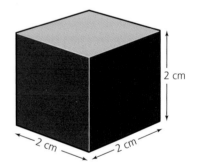

The density of one cube is 0.375 g/cm³

- To calculate the volume of an object, rearrange the equation to:

$$\text{volume} = \frac{\text{mass}}{\text{density}}$$

- To calculate the mass of an object, rearrange the equation to:

mass = density × volume

Know that the unit of density is kg/m³ or g/cm³

Density is measured in kilograms per cubic metre (kg/m^3), or grams per cubic centimetre (g/cm^3).

Understand how to measure density

Regularly shaped solids

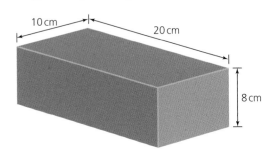

A brick measures 20 cm × 10 cm × 8 cm and has a mass of 2040 g. What is the density of the brick?

Step 1 Find the volume of the brick:

$$\textbf{volume} = \textbf{length} \times \textbf{width} \times \textbf{height}$$
$$= \textbf{20} \times \textbf{10} \times \textbf{8}$$
$$= \textbf{1600 cm}^3$$

Step 2 Calculate the density of the brick:

$$\textbf{density} = \frac{\textbf{mass}}{\textbf{volume}}$$
$$= \frac{\textbf{2040 g}}{\textbf{1600 g/cm}^3}$$
$$= \textbf{1.275 g/cm}^3$$

Irregularly shaped solids

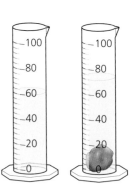

It is not easy to measure the sides of an irregularly shaped object to calculate its volume. Instead, the displacement of water method is used. In the following example, the mass of the stone is 45 g. We need to calculate its density.

Step 1 Fill a measuring cylinder to 60 cm³.

Step 2 Drop in the irregular object and read the new volume.

In this example, the volume is 75 cm³ with the stone in the water.

Step 3 Calculate the volume of the stone:

$$\textbf{75 cm}^3 - \textbf{60 cm}^3 = \textbf{15 cm}^3$$

Step 4 Calculate the density of the stone:

$$\textbf{density} = \frac{\textbf{mass}}{\textbf{volume}}$$
$$= \frac{\textbf{45 g}}{\textbf{15 cm}^3}$$
$$= \textbf{3 g/cm}^3$$

So, the density of the stone is 3 g/cm³

Liquids

Step 1 Measure the mass of an empty measuring cylinder. Record the mass, m_1

Step 2 Pour 50 cm³ (volume) of the liquid into the measuring cylinder and record the new mass, m_2

Step 3 Calculate the mass of the liquid by subtracting $m_2 - m_1$

This figure gives you the mass of 50 cm³ of the liquid.

Step 4 Use these results in the equation above to calculate the density of the liquid.

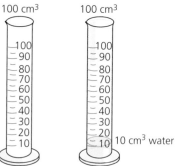

Recognise that air has mass and that it is possible to measure its density

Air also has mass, and it is possible to measure its density using the following method:

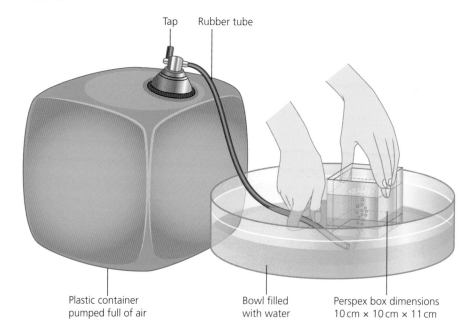

Tap   Rubber tube

Plastic container pumped full of air

Bowl filled with water

Perspex box dimensions 10 cm × 10 cm × 11 cm

Step 1 Measure the mass of the container with its tap open. Record the mass, m_1

Step 2 Pump as much air as possible into the container and close the tap. Measure the mass of the filled container, m_2

Step 3 Immerse the Perspex box into the bowl of water. Invert it to fill it with water, then carefully lift it most of the way out making sure the water stays in place.

Step 4 Place the rubber tube under the box and carefully open the tap on the container filled with air.

Step 5 When the air has replaced 1 litre of water (has gone 10 cm down the Perspex cube), close the tap.

Step 6 Refill the box with water and repeat the process until no more air comes out of the container. Keep a record of how many litres of water are replaced and, if necessary, make an estimate of the volume of air released the last time you try to fill the cube with air. This will give you V, the total volume of air that came out of the container.

Step 7 Calculate the density of the air.
First calculate the mass of the expelled air in the container by subtracting $m_2 - m_1$:

$$\text{density} = \frac{\text{mass}}{\text{volume}}$$

$$= \frac{m_2 - m_1}{V}$$

Recommended practical activities

Recall any investigations you have conducted including:

- measuring the density of different objects, some regular and some irregular
- investigating the effect of density on immiscible liquids of different densities, for example, oil and water
- researching the effect of temperature on the density of water.

Exam-style questions: Exercise 33

1 Which two things about a substance do you need to know to be able to find out its density? (1)

2 Using the words density, mass and volume, write down the equation that you would use to find density. (1)

3 Find the densities of the following blocks of materials:

    a) Material A: mass 750 g, volume 100 cm^3 (1)

    b) Material B: mass 220 g, volume 20 cm^3 (1)

    c) Material C: mass 540 g, volume 200 cm^3 (1)

    d) Material D: mass 162 g, volume 60 cm^3 (1)

4 Which of the blocks A, B, C, D in question 3 are made of the same material? Give a reason. (2)

5 The density of marble is 3.2 g/cm^3; the density of glass is 2.8 g/cm^3. If you had 3 kg of each, which material would have the larger volume? Carry out calculations to check your answer, giving volumes to one decimal place. (3)

6 If the density of air is 1.3 kg/m^3, what is the mass of air in a room measuring 10 m × 6 m × 3 m? (2)

Know that sound is produced by vibrations of objects

Vibrations

● Sounds are made when objects move backwards and forwards or **vibrate**. For example, when a guitar string is plucked, it moves backwards and forwards – it vibrates. When you blow into a recorder or over a bottle, the air inside vibrates.

■ A vibrating guitar string

● **Vibrations** are carried through the air as **sound waves**.
● Vibrations are not always directly visible. Their existence can be demonstrated by, for example, using grains of rice to show how a drumskin moves when it is hit.

Waves

● Sound travels because vibrations are passed from particle to particle in the air. As the particles move backward and forwards, **sound waves** are formed.
● Waves are a way of moving energy from one place to another.
● When sound waves meet our eardrum, it vibrates. In the ear, these vibrations are changed into messages which are sent to the brain, so we hear the sound.

Understand that sound has a different and finite speed in different mediums

● Sound waves are formed when particles vibrate.
● Most of the time, sound waves travel through air. They are also able to travel through solids, liquids and other gases.
● Anything sound can travel through is called a **medium**.
● Sound passes more efficiently through solids and liquids because the particles are closer together and have stronger bonds. In air or other gases, the particles are much further apart (see Chapter 14: States of matter – the particle model).
● There are no particles at all in a **vacuum**. Since there are no particles to vibrate, sound cannot travel through a vacuum.

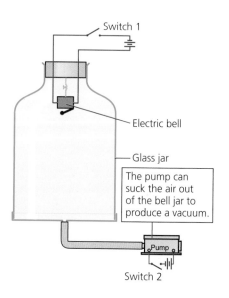

Switch 1

Electric bell

Glass jar

The pump can suck the air out of the bell jar to produce a vacuum.

Pump

Switch 2

Understand the relationship between the loudness of a sound and the amplitude of vibration causing it

Loudness

- Loudness depends upon the **size** of the vibration.
- The size of the vibration is called its **amplitude**.
- The greater the amplitude, the louder the sound.
- This will depend on, for example, how hard a string is plucked or a percussion instrument is hit: the harder something is hit or plucked, the larger the vibration and the louder the sound.

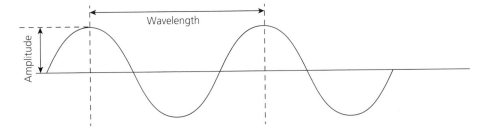

Why sounds become softer the further away you are

- Vibrations send out sound waves in all directions.
- Sound waves spread the energy from the vibrations over a greater area as they travel further from the vibrating source.
- The further away you are from the source, the quieter the sound. The vibrations are smaller, the sound waves have a lower amplitude, so the sound is quieter.
- This is why we see cars on a distant motorway, even though we cannot hear them.

Problems caused by loud sounds

- Loud sounds can perforate the eardrum so it cannot vibrate properly, leading to deafness.
- Louder sounds can damage the inner ear, too.
- If the eardrum is able to repair itself, or the damage to the inner ear is slight, the deafness may be temporary.
- Permanent deafness might occur if the inner ear is badly damaged by very loud sounds.
- Even if loud sounds that damage the ear do not lead to deafness, they can reduce the range of frequencies that can be heard.

■ Loud sounds can cause damage to the eardrum and inner ear

Understand the relationship between the pitch of a sound and the frequency of the vibration causing it

Pitch

- **Pitch** describes how **high or low a sound** is.
- Pitch is determined by:
  - The length of vibrating material: shortening a vibrating string increases the pitch of a sound because a shorter string vibrates more quickly and faster vibrations result in higher-pitched sounds.
  - The amount of material vibrating: using a thicker string lowers the pitch of a sound because thicker strings are heavier, there is more material to move so they vibrate more slowly, resulting in lower-pitched sounds.
  - The tension in a string: tighter strings vibrate more quickly.
- The number of complete vibrations in a specified time (cycles per second) is called the **frequency**.
  - A **faster vibration** has a higher frequency and produces a sound that has a higher note or, more correctly, a note of a **higher pitch**.
  - A **low frequency of vibration** results in a note of **low pitch**.

Frequency

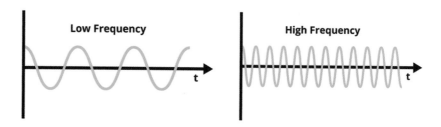

- The **wavelength** is the distance between two successive peaks of a wave. Notes with a higher pitch have a shorter wavelength.

Recognise that humans and animals can hear different ranges of sounds

- The frequency of a sound is measured in Hertz (Hz).
- 1 Hz = 1 cycle per second.
- Different people have different auditory ranges.
  - A young person can hear a wide range of frequencies.
  - In older people, this range is reduced.
- Several animals can hear frequencies well beyond the human hearing range. For example, bats are able to hear sounds that are inaudible to the human ear.

■ A vampire bat

Recommended practical activities

Recall any investigations you have done in class such as:

- measuring the speed of sound by an echo method or by direct measurement using the equation: **speed of sound** = $\dfrac{\text{distance}}{\text{time}}$
- investigating the range of sounds that humans of different ages can hear
- researching the range of sounds that different animals can hear
- researching the effects of loud sounds on hearing.

Exam-style questions: Exercise 34

1 Complete the sentences using words from the list below.
  particle   vibrates   sound waves

  a) Sound is made when something ____ . (1)

  b) Vibrations are carried through the air as ____ . (1)

  c) Sound travels because it is passed from particle to ____ . (1)

2 a) Explain why sound passes more efficiently through solids and liquids than through gases. (1)

  b) Explain why sound does not travel through a vacuum. (1)

3 a) What feature of a sound wave determines:
  i) the pitch of a note? (1)

  ii) loudness of the sound? (1)

  b) Use the features you have described in **a)** to describe two differences between a high, loud note and a low, quiet one. (2)

4 Explain why a dog whistle is virtually inaudible to human ears but seems to work well dog training. (2)

Light is an energy pathway your eyes can detect.

Light sources

The light we see comes from various sources described below.

- **Luminous** sources such as the Sun, a lamp, stars and fire, all emit light.

■ The Sun

■ A camp fire

 Light travels in straight lines from the luminous light source to our eyes.

■ Light from a luminous light source

■ You can often see light that travels in straight lines. Here the sun's rays break through a stormy sky.

- **Reflection** from **non-luminous** objects. Many objects, such as the Moon, a mirror, or a page of white paper, look bright but do not emit light. They are reflecting light from a luminous source into our eyes. The reflected light is scattered from the object and enters the eye.

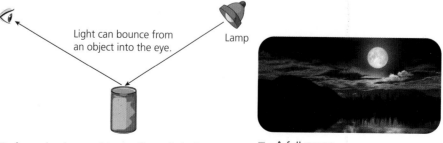

Light can bounce from an object into the eye.

Lamp

■ A non-luminous object reflects light from a luminous source

■ A full moon

How light travels

- Light travels very fast – at a fixed speed in any medium.
- Light rays travel in straight lines – you cannot see an object if there is something blocking these straight lines.
- Light rays travel through some materials but not others:
  - Light rays pass straight through **transparent** materials such as glass, water or clear plastic; you can see clear images through them.

  ▪ Light travels through the transparent glass giving a clear image of the view outside

  - Light rays are scattered as they pass through **translucent** materials such as tracing paper and some plastics, which means that you cannot see a clear image of what is on the other side.

  ▪ When light rays travel through translucent material like this net curtain, they are scattered and you cannot see a clear image of what is outside

  - Light rays cannot pass through **opaque** materials such as wood, metal and pottery at all. These materials block light rays (by absorbing or/and reflecting them to different amounts) and form shadows.

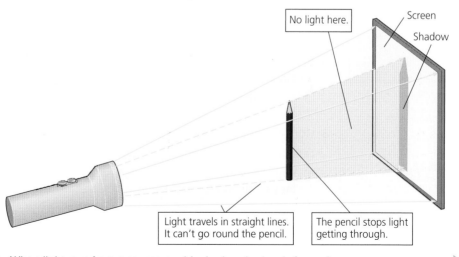

No light here.

Screen

Shadow

Light travels in straight lines. It can't go round the pencil.

The pencil stops light getting through.

▪ When light rays from a source are blocked, a **shadow** is formed

Reflection of light

- Shiny surfaces bounce light rays off the surface at the same angle as they hit the mirror surface. This is called **reflection**.
- Dull surfaces reflect light rays in many directions – the light is **scattered** because these materials are rough, even if only at the level of molecules.

■ The scenery is reflected in the flat 'mirror-like' water of the lake

How plane mirrors reflect light

- A **plane mirror** is one that is flat.
- The **normal** is an imaginary line at 90° to the surface of the mirror that is used for measuring angles.
- Rays coming from a light source to the mirror are called **incident rays** and they hit the mirror at the **angle of incidence**.
- The angle of incidence is the angle between the incident ray and the normal.
- Rays bouncing away from the mirror are called **reflected rays** and they move away at the **angle of reflection**.
- The angle of reflection is the angle between the reflected ray and the normal.

angle of incidence = angle of reflection

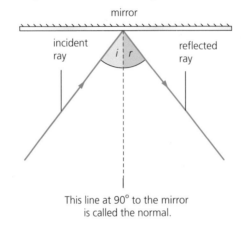

mirror

incident ray i | r reflected ray

i = angle of incidence
r = angle of reflection

This line at 90° to the mirror is called the normal.

Note: When using a protractor to measure these angles, place the baseline of your protractor along the normal.

Test your preliminary knowledge: Exercise 35A

1 A table does not emit light and yet we can see it. Explain how this is possible. **(2)**

2 Write down the words that complete the following sentences.

  a) Light rays will pass through _____ materials and you can see clear images through them. **(1)**

  b) Light rays cannot pass through _____ materials at all. Such materials block light rays and form shadows. **(1)**

  c) Some light rays are scattered as they pass through _____ materials, which means that you cannot see a clear image of what is on the other side. **(1)**

3 Complete the sentences using the words below.

  reflection   incident rays   reflected rays   incidence

  a) Rays coming from a light source to a mirror are called _____ and they hit the mirror at the angle of _____ . **(2)**

  b) Rays bouncing away from a mirror are called _____ and they move away at the angle of. _____ . **(2)**

4 a) What happens when a light ray hits a mirror? **(1)**

  b) If the mirror is a plane mirror, what can you say about the angles you named in question **3**? **(1)**

Know that light can travel through a vacuum, but sound cannot

● Light can travel through a vacuum.
● Sound cannot because there are no particles to transfer the sound when objects vibrate (see Chapter 34: Sound).

There is a famous demonstration of an alarm clock ringing its bell while it is inside a bell jar. As the air is removed by a vacuum pump, the ringing becomes quieter. When all the air has been removed, no sound at all can be heard even though you can see that the hammer on the clock is still hitting the bell (see Chapter 18: Mixtures including solutions).

Know that light travels much faster than sound

● Light travels very fast, much faster than sound.
● We can often see something long before we hear it. For example, during a thunderstorm we see lightening (almost at the same time that it happens) before we hear the thunder.

Understand how and why light is refracted at the boundary between two different materials

● Light travels at different speeds, depending on the medium (material) it is travelling through.
● Light passes easily through gases such as air but moves more slowly through liquids and transparent solids such glass.
● When light rays pass from a less dense medium to a denser medium (for example, from air to water) they slow down and bend towards the normal.
● When light rays pass from a dense medium to a less dense medium (for example, from water to air) they speed up and bend away from the normal.
● The bending of light as it reaches the boundary between different mediums is called **refraction**.

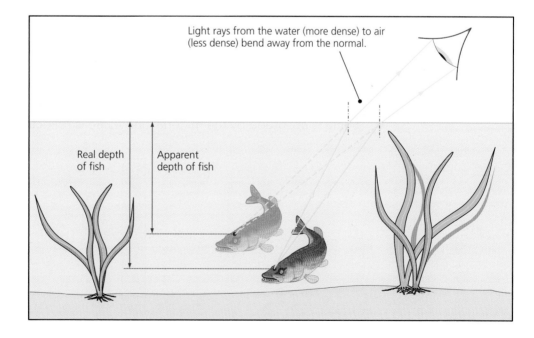

Light rays from the water (more dense) to air (less dense) bend away from the normal.

Real depth of fish

Apparent depth of fish

Understand that light passing through materials can be absorbed or scattered

Light can pass through some materials but not others. We can classify materials depending on how light passes through them (or not).

Transparent materials
- Light passes straight through.
- Light is not scattered and not absorbed.
- You can clearly see an object behind transparent material.

Light rays pass straight through a transparent material.

Translucent materials
- Light passes through but the light rays are changed.
- Some light is absorbed and some scattered in all directions.
- You cannot get a clear picture of an object behind translucent material.

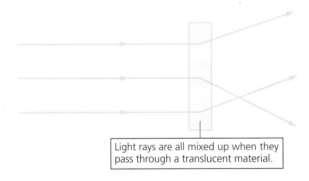

Light rays are all mixed up when they pass through a translucent material.

Opaque materials
- All light is stopped from passing through.
- Some is absorbed by the material and some is reflected.

Recognise that different frequencies of light are refracted differently and that this causes dispersion

- Light is a type of electromagnetic wave and the light we see is just a small part of the electromagnetic spectrum.
- The light we see is called **visible light**.
- **White light** is made up of several different colours, called the visible spectrum. Each colour has a different range of frequencies.
- The splitting of white light into its colours is called **dispersion**.
- An example of dispersion is when white light from the Sun hits a drop of water: the colours are separated producing a rainbow.
- The colours of the rainbow are red, orange, yellow, green, blue, indigo and violet.

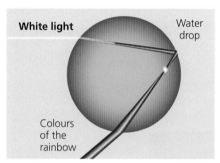

White light

Water drop

Each drop of water splits the white sunlight into a set of different colours.

Colours of the rainbow

Recommended practical activities

Recall any of the following investigations you may have done in class.

1. Use a ray model to show how light is reflected in a mirror.

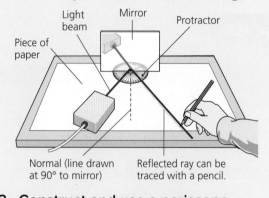

Light beam

Mirror

Protractor

Piece of paper

Normal (line drawn at 90° to mirror)

Reflected ray can be traced with a pencil.

2. Construct and use a periscope.

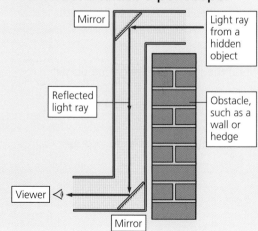

Mirror

Light ray from a hidden object

Reflected light ray

Obstacle, such as a wall or hedge

Viewer

Mirror

A submarine uses a periscope to see what's going on above the water

3. Construct and use a pinhole camera.

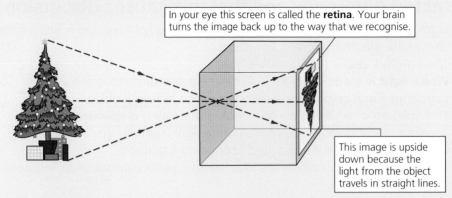

In your eye this screen is called the **retina**. Your brain turns the image back up to the way that we recognise.

This image is upside down because the light from the object travels in straight lines.

■ **A pinhole** camera shows how light travels in straight lines and forms an image on a screen

4. Investigate the dispersion of white light through a prism.

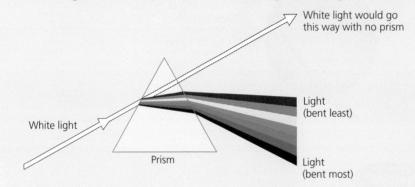

White light would go this way with no prism

Light (bent least)

White light

Prism

Light (bent most)

5. Research the way in which dispersion causes us to see a rainbow.

Exam-style questions: Exercise 35B

1 Light rays are emitted from a luminous source. List three other facts about light rays. (3)

2 a) What is refraction? (1)

b) Where does this happen? (1)

c) Why does this happen? (1)

d) Why is it hard to pick up a coin from the bottom of a glass beaker? Use a diagram to help your explanation. (3)

3 a) Draw a diagram of an instrument that uses two mirrors to allow you to look over a high wall. (4)

b) What is the name of your instrument? (1)

c) Give an example of a situation when the instrument might be useful. (1)

4 a) Use a simple diagram to explain what happens to light when it hits a translucent material. (2)

b) Give an example of a translucent material you might find at home. (1)

5 Describe how rainbows are formed. (2)

Preliminary knowledge

These are the electrical symbols for **components** (parts of a **circuit**) which you are expected to know for Common Entrance papers.

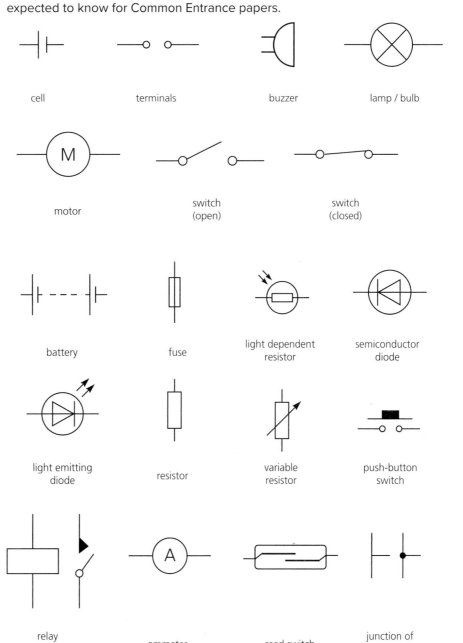

cell   terminals   buzzer   lamp / bulb

motor   switch (open)   switch (closed)

battery   fuse   light dependent resistor   semiconductor diode

light emitting diode   resistor   variable resistor   push-button switch

relay (normally open)   ammeter   reed switch   junction of conductors

Simple electrical circuits

- Electricity flows around a circuit. The flowing electricity is called an **electrical current**.
- Electricity transfers energy from a chemical store of energy in the **cell** (or the battery of several cells) to components in the circuit.
- Electricity flows through **wires**. These wires are electrical conductors.
- For an electrical circuit to work:
  - there must be no gaps between wires: it must be a **complete circuit**
  - all cells must face the same way.

Drawing electrical circuits

- Circuits should be drawn using straight lines for the conducting wires.
- Circuit symbols should be used to represent the various components.

Here are some examples of electrical circuits.

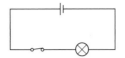

Brighter/normal/dimmer

- A switch, a lamp and a cell in series
- A switch, two lamps and two cells in series
- Toy car: a switch, two lamps, two cells, a motor and a buzzer

Know the function of and symbols for common devices

Component	Symbol	Details	Examples
SPST switch (closed)		When closed, the circuit is complete and the current (electricity) flows.	
SPST switch (open)		When open, the circuit is broken and no current (electricity) flows.	
push-button switch		When pressed, the circuit is completed, allowing current to flow.	bell push, computer keys
reed switch		Reed switches are normally open and need a magnet to close them.	door sensors in alarm systems, fluid-level sensors in dishwashers
resistor		Used to control or limit the flow of current (electricity) in a circuit.	
variable resistor		Has a sliding contact that moves along a conductor (often a coil of wire) so you can change the length of conductor that is included in the circuit and so vary the amount of current that is able to flow.	a dimmer switch controlling room lighting

Component	Symbol	Details	Examples
motor	(M)	A device that uses electricity to transfer energy from a chemical store to a kinetic (movement) store.	food mixer, hair dryer
buzzer		A device that transfers energy by sound to a kinetic store in air particles, making them vibrate. We detect this as a sound.	doorbell
LDR (light-dependent resistor)		A component with a resistance that depends on light intensity: usually the resistance is low in bright light and high in the dark.	automatic security lights, driving lights on cars
LED (light-emitting diode)		A compact light source that emits light when a small current flows through it.	flashing cycle lights, 'on' indicator light on a television
fuse		Too much current can damage a device so a fuse breaks a circuit if the current in the wire becomes too great.	in a plug

Understand the difference between a parallel and series circuit

Series circuit

- One path for the current to travel. There are no branches or junctions.
- Components are connected one after the other – rather like railway carriages.
- Current is the same everywhere in a **series circuit**.
- Every component has to work or none of them will work.

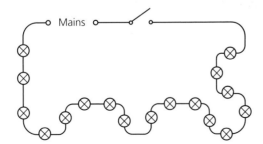

Series circuit: Christmas tree lights

Parallel circuit

- **Parallel circuits** can be thought of as two or more individual circuits connected to the same power supply.
- Each branch draws its own supply of current from the cell/battery.
- Adding extra components in parallel means more current is drawn so the battery/cell runs down more quickly.
- If one component fails, components in other circuits are not affected.

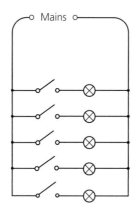

■ Parallel circuit: household lights

Recognise that the current in a series circuit depends on the number of cells and the number and nature of other components

When a lamp, cell and switch are connected one after the other in series and when there is only one path for the current to follow, we can observe the following:

- The lamp in this circuit is said to shine with 'normal brightness'.
- Adding more cells in series increases the current that will be pushed round the circuit. This makes the lamp in the circuit brighter.
- Adding more lamps or other components in series will make it more difficult for the current to flow. The lamp or lamps will be dimmer.

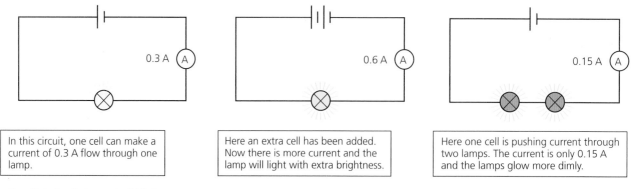

In this circuit, one cell can make a current of 0.3 A flow through one lamp.

Here an extra cell has been added. Now there is more current and the lamp will light with extra brightness.

Here one cell is pushing current through two lamps. The current is only 0.15 A and the lamps glow more dimly.

■ Parallel circuit: household lights

The symbol —Ⓐ— represents an instrument called an ammeter, used for measuring the electrical current in a circuit (see Chapter 37: Currents).

Recommended practical activity

Recall how you may have constructed series and parallel circuits from circuit diagrams.

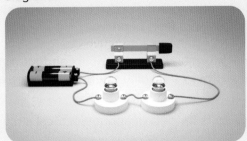

■ A simple series circuit

■ A simple parallel circuit

Exam-style questions: Exercise 36

1 In a circuit with one cell and one lamp, the lamp glows with normal brightness. Copy and complete the sentences below using phrases from the list.

more brightly   **with normal brightness**   **only dimly**

a) If there are two cells and two lamps, the lamps will shine ____ .   (1)

b) If there are three cells and two lamps, the lamps will shine ____ .   (1)

c) If there is one cell and two lamps, the lamps will shine ____ .   (1)

2 a) Draw a series circuit with two cells, three lamps and a switch.   (3)

b) In a circuit with one cell and one lamp, the lamp glows with normal brightness. How brightly will the lamps in the circuit you have drawn shine?   (1)

c) When the switch is closed, the lamps do not light. The cells are not flat. Give three possible reasons why the lamps do not light.   (3)

3 Match each symbol with its description.   (5)

Symbol		Details
A	⊏▭⊐	**1** Lamp
B	⊕	**2** Junction for electrical wires
C	⊗	**3** Needs a magnet to close it
D	┤•	**4** A chemical store of energy
E	⊣⊢ - - ⊣⊢	**5** Gives a very bright light when only a small current flows

4 Choose words from the list below to complete the sentences.

parallel   **series**   **the same**   **dimmer**   **different**   **brighter**

a) In a series circuit, the current is ____ in different places.   (1)

b) If one component in a ____ circuit fails, the other components are unaffected.   (1)

c) If one component in a ____ circuit fails, nothing else in the circuit will work.   (1)

d) If more components are added in series to a circuit that includes two lamps, the lamps will become ____ .   (1)

- The flow of electricity around a circuit is called the **electrical current**.
- An electrical current is a **flow of charge**.
- When electrons (which have a negative charge) flow, they set up a current.
- Electricity flows through wires (**electrical conductors**).
- **Electricity is a pathway** by which energy is transferred from one store to another.

(See Chapter 36: Circuits for details about electrical circuits.)

Know how to measure current and that its unit is the ampere

- To measure the electrical current in a circuit, you use an instrument called an **ammeter**, which is connected in series within the circuit.
- The size of the current is measured in units called **amperes** (A) which is often shortened to **amps**.

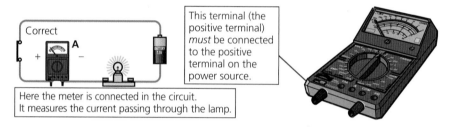

Correct

This terminal (the positive terminal) *must* be connected to the positive terminal on the power source.

Here the meter is connected in the circuit.
It measures the current passing through the lamp.

- Remember that the symbol —Ⓐ— is used to represent an ammeter in circuit diagrams.
- If more bulbs or components are added in series to a circuit it will be more difficult for the current to flow. Any bulbs will be dimmer and the ammeter will give a **lower** reading.

Understand current as flow of charge that is not 'used up' by components

- Using an ammeter in a circuit you can demonstrate that the current passes through a component but is not 'used up'.
- In the example below, the reading (0.6 A) is the same at each point in the circuit.

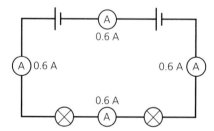

- The current is the flow of charge that leaves the power source, goes through the components and back to the power source.

Recognise differences in resistance between conducting and insulating components (qualitative only)

- Current flows because the power supply provides a voltage to 'push' the electric charge.

Good conductors

- Components that allow electricity to **flow easily** are **good conductors**.
- They have a **low resistance**.
- Wires in the home, for example, are made of copper, which is a good conductor.
- Thicker wires are better conductors than thinner wires.

Good insulators

- Components that make it **hard** for the current to flow are **good insulators**.
- They have very **high resistance**.

Controlling the flow

- A resistor or variable resistor can be used to control the flow of a current.
- It can protect components such as lamps from blowing.
- It can be used to reduce the current flowing through a component to, for example, dim a lamp.

■ Lamp with dimmer switch

Recommended practical activity

Recall any practical activities you may have undertaken to:

- measure the current in different parts of circuits using an ammeter
- investigate the resistance of different materials using a series circuit and ammeter
- research practical circuits for devices like alarms, light gauges, and so on.

Exam-style questions: Exercise 37

1 Choose the word from each list that best completes the sentence above it:

a) An instrument that can be used to measure the electrical current in a circuit is called an _____ . (1)

resistor   fuse   ammeter   ampere

b) The unit of electrical current is the _____ . (1)

newton   amp   gram   hertz

c) Components that allow electricity to flow easily are good _____ . (1)

insulators   conductors   switches   cells

d) Good conductors have _____ resistance. (1)

low   high   no

e) Electricity is a pathway by which _____ is transferred from one store to another. (1)

heat   energy   light   sound

2 a) Draw a circuit that contains two cells, a lamp and an ammeter in series. (2)

b) What happens to the ammeter reading if another lamp is added to the circuit in series with the first? (1)

c) What must you be careful about when connecting the ammeter? (1)

3 Look at the circuit below.

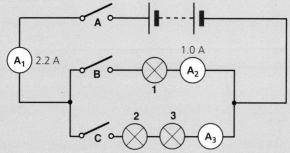

a) Which switch or switches would you use to:

i) control all lamps? (1)

ii) control lamp 1 only? (1)

b) Which switch or switches would you close to light:

i) lamp 1 only? (1)

ii) lamps 2 and 3 only? (1)

c) A current of 2.2 A flows through ammeter A_1 when switches A, B and C are closed.

i) If ammeter A_2 has a reading of 1.0 A when switches A, B and C are closed, state what the reading on ammeter A_3 will be. (1)

ii) State what the reading on ammeter A_3 will be if switches A and C are closed and switch B is open. (1)

Preliminary knowledge

Magnetic forces

- A force is a push or pull – you can only see what a force does.
- One type of force is a **magnetic force**.
- A freely suspended **magnet** will always have the same end pointing towards the magnetic north pole of the Earth, so this end is called the north pole of the magnet. The other end is called the south pole of the magnet. The freely suspended bar magnet acts as a compass.
- At present, the Earth's magnetic north pole is close to the geographical North Pole.
- A compass needle is a small magnet.
- A magnetic force can work at a distance (as it does when, for example, in the action of a compass needle) whereas pushing a door open requires direct contact.

Forces of attraction and repulsion

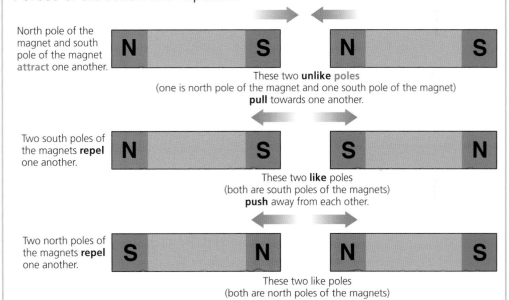

North pole of the magnet and south pole of the magnet **attract** one another.

These two **unlike poles**
(one is north pole of the magnet and one south pole of the magnet)
pull towards one another.

Two south poles of the magnets **repel** one another.

These two **like** poles
(both are south poles of the magnets)
push away from each other.

Two north poles of the magnets **repel** one another.

These two like poles
(both are north poles of the magnets)
push away from each other.

Magnetic materials

- A **magnetic material or object** is one that will always be **attracted** to a magnet.
- The only **magnetic elements** are iron, nickel and cobalt.
- Alloys that contain any of these elements are also magnetic; for example, steel is magnetic.
- Most substances are not magnetic.

Test your preliminary knowledge: Exercise 38A

1 Copy and complete the following sentences.

   a) Materials that are attracted by a magnet are said to be ＿＿＿ . (1)

   b) The end of a magnet that points towards the geographical North Pole is called the ＿＿＿ . (1)

   c) ＿＿＿ poles attract; ＿＿＿ poles repel. (2)

Know about magnetic fields as regions of space where magnetic materials experience forces

- A **magnetic field** is the area around a magnet where a magnetic force can be detected.
- You cannot see the magnetic fields but you can see them using iron filings (see 'Recommended practical activities' at the end of the chapter).
- When you bring a metal object near a magnet, you can feel the pull before the magnet and object touch.
- The magnetism of the magnet extends into the air beyond it.
- We know magnetism is a force because it can push and pull another object.

Understand how to represent a magnetic field using field lines with arrows

When you draw a magnetic field:

- Draw the lines that show the lines of magnetic force.
- Draw arrows that point away from the direction on which the force on a north pole acts. These arrows show the direction in which the force acts.
- The spacing of the lines tells us the strength of the field. Lines closer together show a stronger field than lines further apart.

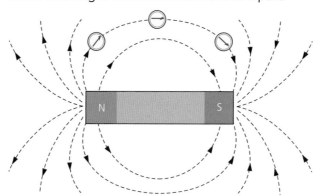

Understand that the Earth has a magnetic field

The Earth has a magnetic field that you can 'see' using a compass. The needle of a compass is a magnet that always points towards the Earth's north pole. It does this because the Earth behaves like a giant magnet. We can use this to find out which way to travel.

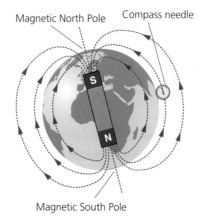

Magnetic North Pole   Compass needle

Magnetic South Pole

■ The Earth behaves like a giant magnet with a south pole of the compass at the magnetic north pole

Recognise that repulsion by a known magnet is the only true test for another magnet

If something is attracted to a magnet, it could be a magnetic material or another magnet. This means that the only way to test if a material or object is a magnet is to see if it can be repelled by a magnet.

Know that a current in a coil produces a magnetic field

An electrical current can be used to make magnets.

Making an electromagnet

- When an electric current flows through a wire, a magnetic field is produced around the wire (see 'Recommended practical activities' at the end of the chapter).
- A compass needle shows that this magnetic field has a direction.
- If the direction of the electrical current is changed, the needle points in the opposite direction.
- If the wire is formed into a loop, the magnetic field through the loop will be in one direction at right angles to it.
- If more loops are added using insulated wire to make a coil, there is a strong magnetic field inside the coil that:
    - ceases if the current stops flowing
    - is in one particular direction
    - will reverse in direction if the direction of the current is reversed.
- A soft iron rod (core) placed inside the coil will become a magnet when the current is switched on.

- To make an electromagnet stronger:
    - add more turns of insulated wire to the coil
    - increase the current
    - add a 'core' of soft iron.

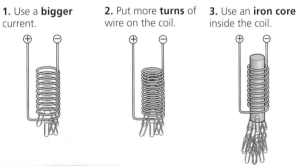

1. Use a **bigger** current.　　**2.** Put more **turns** of wire on the coil.　　**3.** Use an **iron core** inside the coil.

■ An electromagnet being used to attract magnetic iron filings

Recognise that electromagnets have a wide range of uses

Electromagnets have a wide range of uses because they can be switched on and off.

In scrapyards

- Iron and steel can be separated from non-magnetic materials using an electromagnet.
- Iron can be lifted and moved when the electromagnet is on, and then dropped where it is needed by turning off the current.

Relays – electrically controlled switches

- A relay is a switch that is operated by an electromagnet. There are two separate circuits:
  - In circuit 1 the ON/OFF switch controls a small current that is used to make a coil into a magnet. This is positioned so it can pull the switch in circuit 2 closed when the coil is a magnet.
  - Circuit 2 is the 'end use' circuit which often has a much larger current.

The starter motor in a petrol or diesel vehicle uses a relay. (Note that electric vehicles do not have starter motors!)

Other uses

- Maglev (from 'magnetic levitation') trains use electromagnets to float above the track they move along.
- Electric bells.
- Reed switches in electronic circuits (see Chapter 36: Circuits).

■ An electromagnet being used to separate iron and steel from non-magnetic material

■ A Maglev (magnetic levitation) train

Recommended practical activities

1. **Investigate the field of a bar magnet and an electromagnet using a plotting compass or iron filings.**

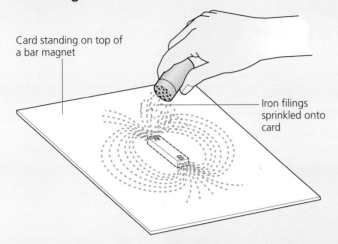

Card standing on top of a bar magnet

Iron filings sprinkled onto card

■ Investigating the magnetic field of a bar magnet. Lines of magnetic fields are invisible – we only see them because of the use of iron filings.

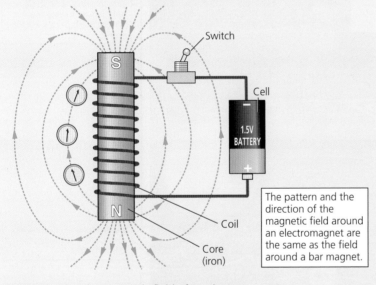

Switch

Cell

1.5V BATTERY

Coil

Core (iron)

The pattern and the direction of the magnetic field around an electromagnet are the same as the field around a bar magnet.

■ Investigating the magnetic field of an electromagnet

2. **Investigate electromagnets:**
   - Investigate what affects the strength of an electromagnet.
   - Research the uses of electromagnets.

Exam-style questions: Exercise 38B

1. Copy the diagram below and draw the magnetic field around the magnet. (2)

N   S

2. a) How would you use a cell and some insulated wire to magnetise an iron nail? (3)

   b) How would you test the iron nail to show that it had become a temporary magnet? (2)

   c) What could you do to make the nail a stronger magnet? (2)

The relative positions of the Earth, Sun and planets in the solar system

- The nearest star to the Earth is the Sun.
- A planet is a body that orbits a star.
- The Earth, Sun and all the planets are approximately spherical in shape.
- The Earth is one of eight major planets that, along with the dwarf planet Pluto and many asteroids, orbit the Sun making what is known as the **solar system**.

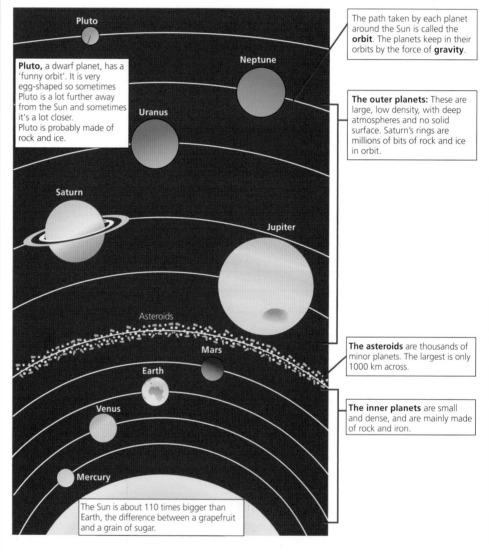

Pluto

The path taken by each planet around the Sun is called the **orbit**. The planets keep in their orbits by the force of **gravity**.

Pluto, a dwarf planet, has a 'funny orbit'. It is very egg-shaped so sometimes Pluto is a lot further away from the Sun and sometimes it's a lot closer.
Pluto is probably made of rock and ice.

Neptune

Uranus

The outer planets: These are large, low density, with deep atmospheres and no solid surface. Saturn's rings are millions of bits of rock and ice in orbit.

Saturn

Jupiter

Asteroids

The asteroids are thousands of minor planets. The largest is only 1000 km across.

Mars

Earth

The inner planets are small and dense, and are mainly made of rock and iron.

Venus

Mercury

The Sun is about 110 times bigger than Earth, the difference between a grapefruit and a grain of sugar.

- To help you remember the order of the planets, use the phrase:
  My **V**ery **E**ccentric **M**other **J**ust **S**aw **U**ncle **N**orman
- The Earth is kept in the Sun's orbit by the pull of the Sun's gravity.

The Moon

- The general name given to a body that orbits another body is a **satellite**.
- Each of the planets, except Mercury and Venus, has at least one satellite, called **a moon**.
- The Earth has one natural satellite (**the Moon**) and this orbits the Earth once every **27 days**. This period of time is called a **lunar month**.
- The Moon is the Earth's nearest neighbour.
- The Moon rotates once during one orbit of the Earth.
- The Moon is not a light source and it is hard to see during daytime. We see it best at night because light from the Sun is reflected from it.
- We do not always see the whole Moon all the time: it appears to change shape at different times of the month. We call these changes in the shape of the Moon **phases**.

Sun and shadows

- The Sun appears to rise in the east, become high in the sky to the south at midday and appears to set in the west. In fact, the Sun does not move, it is the spinning of the Earth that makes it seem to move.
- Opaque objects that block out sunlight cause shadows. These vary in length and direction throughout the day.
- Shadows are always on the opposite side of the object from the Sun.

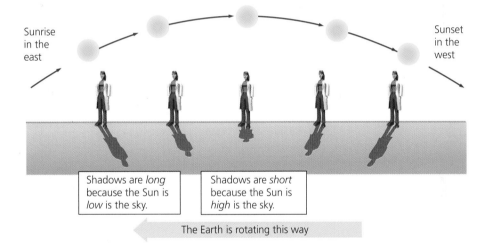

Sunrise in the east

Sunset in the west

Shadows are *long* because the Sun is *low* is the sky.

Shadows are *short* because the Sun is *high* is the sky.

The Earth is rotating this way

- The size and direction of a shadow depends where the Sun is relative to the object it is shining on

Test your preliminary knowledge: Exercise 39A

1 Why do we experience night and day? (2)

2 What causes the Sun to appear to move across the sky? (1)

3 Explain the difference in the appearance of shadows cast at midday and in the evening. (2)

4 How long is a lunar month? What does this period of time represent? (2)

5 Why can we see the Moon better at night than in the daytime? (2)

Know how the movement of the Earth causes the apparent daily and annual movement of the Sun and other stars

Day and night

- The Earth spins on its own axis and makes one complete turn every **24 hours** – this is called a **day**.
- As the Earth spins:
  - half of it is lit up by the Sun – **daytime**
  - the other half is in darkness – **night-time**.

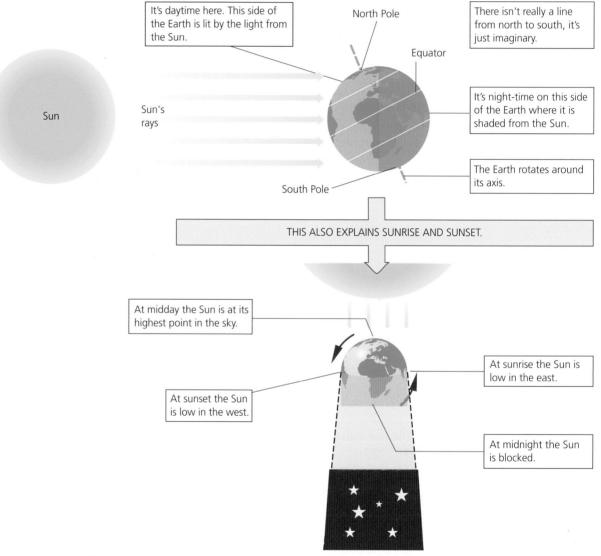

It's daytime here. This side of the Earth is lit by the light from the Sun.

North Pole

There isn't really a line from north to south, it's just imaginary.

Equator

Sun

Sun's rays

It's night-time on this side of the Earth where it is shaded from the Sun.

The Earth rotates around its axis.

South Pole

THIS ALSO EXPLAINS SUNRISE AND SUNSET.

At midday the Sun is at its highest point in the sky.

At sunrise the Sun is low in the east.

At sunset the Sun is low in the west.

At midnight the Sun is blocked.

■ Night and day

In a year

- The Earth takes **365¼ days** to orbit the Sun – this is called **a year**.
- The annual movement of the Earth around the Sun affects seasons (see the table on the following page).

Know about the seasons and the Earth's tilt, and how this affects day length at different times of year, in different hemispheres

The Earth's axis is not vertical but is tilted by about 23°. This means that at any one time, part of the Earth's surface is closer to the Sun than at other times, giving us seasons.

Season	Height of Sun	Temperature of Earth's surface	Length of shadow	Length of day
Summer	High	Warm	Short	Long
Winter	Low	Cold	Long	Short

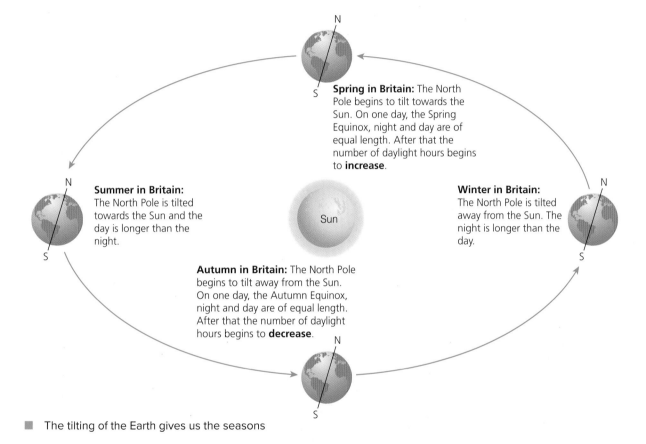

Spring in Britain: The North Pole begins to tilt towards the Sun. On one day, the Spring Equinox, night and day are of equal length. After that the number of daylight hours begins to **increase**.

Summer in Britain: The North Pole is tilted towards the Sun and the day is longer than the night.

Winter in Britain: The North Pole is tilted away from the Sun. The night is longer than the day.

Autumn in Britain: The North Pole begins to tilt away from the Sun. On one day, the Autumn Equinox, night and day are of equal length. After that the number of daylight hours begins to **decrease**.

■ The tilting of the Earth gives us the seasons

Understand that the Earth is one of several planets which orbit the Sun

- Within the **universe** there are more than **100 billion galaxies**.
- A **galaxy** is the name given to a group of stars. A galaxy contains over 100 000 stars. Our galaxy (called the **Milky Way**) is estimated to contain 100 billion stars.
- The Milky Way is so huge, it takes light 100 000 years to cross it. We say that the distance across the Milky Way is 100 000 **light years**.
- A **light year** is the distance that light (travelling at 300 000 km/s) travels in one year. (It is 9 000 000 000 000 km or about 60 000 times greater than the distance from the Earth to the Sun.)
- Our nearest star (called Alpha Centauri) is about 4.2 light years away.
- We can see stars because they are luminous sources and give out light. We see planets because they reflect light from the Sun.
- The planets in our solar system are held in their orbits by gravitational attraction between them and the Sun.

Understand that the positions of the Moon and Earth relative to the Sun can cause solar or lunar eclipses

Lunar eclipse (eclipse of the Moon)

This happens when the Earth is between the Moon and the Sun.

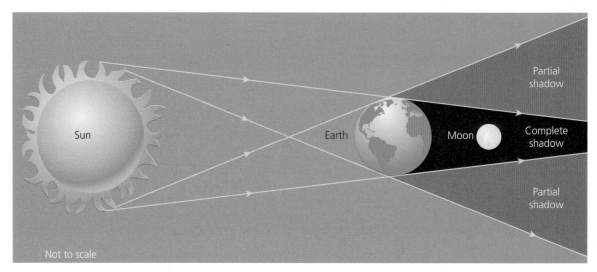

A lunar eclipse

Solar eclipse (eclipse of the Sun)

This happens when the Moon is between the Earth and the Sun.

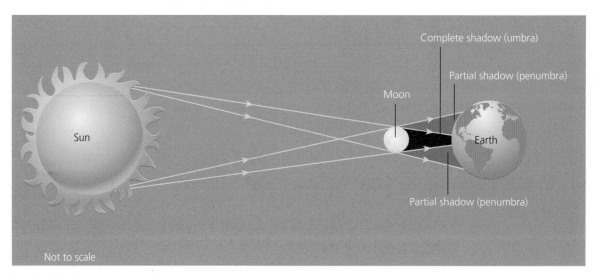

A solar eclipse

Recognise that artificial satellites and probes are used to observe the Earth and to explore the solar system

Natural satellites

A satellite is something that orbits a larger body. Their speed and the gravitational force acting on them keeps them in orbit.

Natural satellites in the solar system include:

- planets orbiting the Sun
- the Moon orbiting the Earth.

Artificial satellites

Artificial satellites are human-made and are launched into space for a particular purpose. These include:

- global positioning (gps)
- observation of the Earth for weather forecasting; climate monitoring; geological surveys, and so on
- observation for military purposes
- communications: satellite TV, radio, mobile phones
- observing space without the interference of the atmosphere: the Hubble telescope is, perhaps, the most famous orbiting telescope.

Recommended practical activities

1. Research online resources for images of the Moon and planets in the solar system taken with telescopes.

2. Research online resources for images of solar or lunar eclipses.

■ Examples of a partial lunar eclipse, as the Earth passes between the Sun and the Moon

■ A lunar eclipse

■ A solar eclipse, as the Moon passes between the Sun and the Earth

3. Research the use of artificial satellites for observations, global positioning and communications.

■ Space satellite over the Earth

■ GPS (global positioning system) receiving location data from artificial satellites orbiting the Earth

Exam-style questions: Exercise 39B

1 What causes us to have seasons on Earth? (2)

2 Write the words that complete the following sentences.

   a) An eclipse of the Moon (lunar eclipse) happens when the _____ is between the Moon and the _____ . (2)

   b) An eclipse of the Sun (solar eclipse) happens when the _____ is between the _____ and the Sun. (2)

3 What does the term light year refer to? (1)

4 The Hubble telescope operates from a satellite. Why should this provide better pictures of space than land-based telescopes? (2)

Ask yourself

What would happen if all satellites stopped working?

Physics – Test yourself

Before moving on to the next chapter, make sure you can answer the following questions. The answers are at the back of the book.

1 Suggest what apparatus and units you would use to measure the following:
   a) The height of a Bunsen burner.
   b) The mass of an apple.
   c) The volume of a wooden pencil box.
   d) The volume of water left in your water bottle.
   e) The volume of a small bunch of keys.

2 Drawing pins come in small boxes, each measuring 2 cm × 3 cm × 0.5 cm. How many boxes of drawing pins could you empty into a large box measuring 60 cm × 30 cm × 15 cm?

3 The density of water is 1 g/cm³. What is the mass of water in a box that measures 30 cm × 50 cm × 20 cm?

4 Draw a circuit that has one cell and one switch, which will light two lamps to 'normal' brightness.

5 Write down the equation used to calculate the weight of an object. Include all units.

6 A spring is 6 cm long. When a load of 100 g is attached to it, the new length is 8 cm. It returns to 6 cm when the load is removed. What will be the length when:
   a) a load of 50 g is attached?
   b) a load of 75 g is attached?

7 A ray of light hits a mirror at an angle.
   a) What is the name of this ray?
   b) What happens to it when it hits the mirror? (Use the scientific term.)
   c) Describe the direction the ray travels in after it leaves the mirror.

8 Draw a diagram to show how you would split a ray of light that is a mixture of red and blue light, into its separate colours.

9 Complete the following table.

Season	Height of Sun	Length of shadow	Length of day
Summer			
Winter			

10 Put the following in ascending order of size:
   star   universe   planet   solar system   galaxy

11 What is energy?

12 List five different stores of energy.

13 a) Name three effects of a force.
   b) Name four types of force.
   c) What is the unit used to measure force?
   d) i) What happens when an unbalanced force acts on an object?
      ii) Give an example of a situation when forces are unbalanced.

Answers

Biology

1 The organisation of living organisms

Exercise 1 (page 11)

1 **a)** cell → tissue → organ → system   (1)

      b) any suitable answers for each component, for
            example: cell – cheek cell, nerve cell; tissue – muscle
            tissue, nerve tissue; organ – liver, lungs; system – circulatory
            system, nervous system.   (4)

2 **a)** stomach   (1)

      b) heart   (1)

      c) brain   (1)

      d) intestines   (1)

      e) lungs   (1)

      f) liver   (1)

3 Pupils should mention the following life processes: (6)

- Cells combine to make tissues and the tissues combine to form an organ, in this case skin.
- Every cell in the skin needs to carry out these life processes to survive:
  - Respiration: transfers energy from the chemical store of energy in glucose to the energy stores of the cell. Needed for growth and repair.
  - Sensitivity: skin needs to be sensitive to things that may cause it damage, e.g. heat.
  - Growth: skin cells need to increase in cell number and size to deal with repairs or to accommodate the growth in size of the body.
  - Nutrition: the process of digestion breaks down food and the energy released is transferred to the cells for growth and repair.
  - Excretion: waste products from the process of respiration need to be removed from the body.

2 Comparing plant and animal cells

Exercise 2 (page 14)

1 A3, B1, C6, D4, E5, F2   (6)

2   (4)

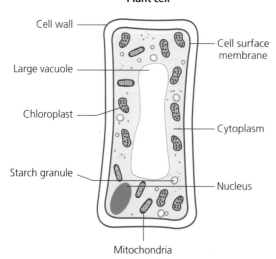

Plant cell

Cell wall — Cell surface membrane — Large vacuole — Chloroplast — Cytoplasm — Starch granule — Nucleus — Mitochondria

3 No. Root cells underground receive no light and do not carry out photosynthesis.   (2)

4   ×500   (10 × 50)   (1)
5 **a)** Plant cell   (1)
    b) Starch cells have been stained black by iodine   (1)
    c) Any four from: cell wall, vacuole, chloroplasts, cell surface
        membrane, cytoplasm, nucleus, mitochondria.   (4)

3 Food nutrients

Exercise 3A (page 15)

1   A3, B4, C2, D1   (4)
2 **a)** movement, respiration, sensitivity, growth, reproduction, excretion
        and nutrition (in any order)   (7)
    b) Oxygen, respiration, chemical, cells   (4)
    c) sugar   (1)
    d) Iodine, blue/black/blue-black   (2)

Exercise 3B (page 17)

1   A4, B6, C5, D1, E2, F3   (6)
2 **a)** meat or cheese   (1)
    b) The emulsion test: mix substance with 95% ethanol, add an
        equal volume of distilled water.
        If fat present, a milky-white emulsion forms.   (2)
3   Vegetables provide mineral salts, vitamins, dietary fibre and water.   (4)

4 A healthy diet

Exercise 4 (page 20)

1 **a)** develop strong teeth   (1)
    b) fibre   (1)
    c) processed foods   (1)
    d) scurvy   (1)
2 **a)** Any one from: bread, bacon, butter, orange juice.   (1)
    b) bacon, egg   (2)
    c) **i)**   sugar
        ii) starch   (2)
    d) Any two from: water, vitamins, minerals, fibre.   (2)
3   At 18, protein will be required mainly for growth.
    At 45, growth is completed, so protein is required for repair and
    replacement of worn-out cells.   (2)
4 **a)** **i)**   Y   (1)
        ii) Although it is mainly indigestible, it provides bulk to enable
            food to pass through the digestive system more efficiently.   (1)
    b) **i)**   Snack X   (1)
        ii) Because it contains the most carbohydrate, a good store of
            energy.   (1)
5   Any two of the following (1 mark for name, 1 for use):   (4)
- vitamins/vitamin C (tissue repair and disease resistance)
- vitamin A (vital for good eyesight, healthy skin)
- vitamin K (helps with blood clotting)
- minerals/ iron (aiding the manufacture of red blood cells/preventing anaemia)
- calcium (supporting good development of bones and teeth, blood clotting)
- salt (to balance fluids in the body for healthy blood pressure)

Ask yourself (page 21)

Think about the types of food needed for a balanced diet: carbohydrates, fats, proteins, dietary fibre, water, minerals and vitamins.

- What nutrients are missing from the diet of the Madagascan people?
  - Little protein to be obtained from plants and animals: lack of water will mean there is little plant life and little for animals to consume.
  - Lack of water to drink.
  - Few carbohydrates available because plants such as grains, fruit and other food crops will struggle to grow.
  - Small amounts of dietary fibre may be obtained from the plants they do manage to find.
- How might this affect their health?
  - dehydration
  - poor growth and development
  - starvation
  - constipation
  - susceptibility to diseases
  - anaemia
  - eventual death

5 Breathing in humans

Exercise 5A (page 22)

1 a) Oxygen, nose, lungs (3)
   b) bloodstream, oxygen (2)
   c) lungs, carbon dioxide (2)
2 The lung surface is greatly folded creating a large surface area for gaseous exchange. (1)
3 Tar in tobacco smoke covers the surface of lungs, reducing the surface area across which gases can be exchanged. (1)

Exercise 5B (page 26)

1 a) air (1)
   b) out of (1)
   c) air sacs (1)
   d) Emphysema (1)
   e) intercostal (1)
2 a) air sacs (1)
   b) i) oxygen enters the bloodstream (1)
      ii) carbon dioxide leaves the bloodstream (1)
3 We breathe air in and out. Air contains oxygen, carbon dioxide and nitrogen. It is the proportion of oxygen and carbon dioxide that changes. Nitrogen content stays the same because it plays no part in respiration. (2)
4 Pass the exhaled air through limewater. If carbon dioxide is present, the limewater will turn milky. (1)

Ask yourself (page 26)

- Particles (e.g. diesel particulates) in polluted air can cause damage to the lining of the lungs.
- This reduces the efficiency of the lungs. The amount of oxygen able to enter the bloodstream drops because the surface of the lungs over which gaseous exchange can take place is reduced.
- It can cause breathing problems such as asthma, emphysema and cancer.

6 Reproduction in humans

Exercise 6A (page 28)

1 A3, B4, C2, D1 (4)
2 **a)** 21 days (1)
   b) cat (1)
   c) 280 days (1)
   d) horse (1)
   e) elephant (1)

Exercise 6B (page 31)

1 **a)** A: prostate gland; B: sperm duct; C: urethra; D: testis/testicle;
      E: scrotum/scrotal sac; F: penis (3)
   b) Prostate gland produces the seminal fluid (1)
      Sperm duct carries sperm from testis to the urethra (1)
      Urethra carries semen from the sperm duct to tip of penis. It also
      carries urine (1)
      Testes are made up of coiled tubes that produce sperm (1)
      Scrotum encloses testes (1)
      Penis carries urine or semen out of the body. It becomes erect during
      sexual intercourse to allow sperm to be released into the female (1)
2 **a)** A: oviduct or fallopian tube; B: ovary; C: muscular wall of the uterus;
      D: cervix; E: vagina; F: opening of vagina (3)
   b) Oviduct or fallopian tube carries egg towards the uterus; (1)
      Ovary contains follicles that develop into eggs. It also produces the
      female hormones oestrogen and progesterone (1)
      Muscular wall of the uterus contracts during childbirth (1)
      Cervix is the neck of the uterus (1)
      Vagina receives the penis during intercourse and is the way out for
      the baby at birth (1)
      Opening of vagina is where penis enters or baby exits (1)
3 **a)** in the oviduct (1)
   b) **i)** the lining of the uterus breaks down (1)
      ii) menstruation (1)
      iii) menstrual cycle (1)
      iv) 28 days (1)
4 **a)** waste materials such as carbon dioxide (1)
   b) soluble materials, such as nutrients (glucose and minerals)
      oxygen for aerobic respiration (2)
   c) **i)** Any two from: alcohol, chemicals from smoking, nicotine, viruses. (2)
      ii) Effects matching two substances selected:
         ● alcohol: brain damage
         ● chemicals from smoking: low birth weight
         ● nicotine: baby can be a nicotine addict at birth
         ● viruses: baby can become infected (2)

Ask yourself (page 32)

The pregnant mother is bitten by a mosquito carrying the Zika virus. The virus
passes into the mother's bloodstream. The virus then crosses from the mother to
her developing baby via the placenta. The virus can then infect the baby causing
microcephaly.

7 Photosynthesis

Exercise 7A (page 34)

1 1: flower; 2: leaf; 3: stem; 4: root (4)
2 A4, B1, C2, D3 (4)

3   a)   Light   (1)
    b)   water, food   (1)
    c)   water, carbon dioxide   (1)
    d)   Flowers   (1)

Exercise 7B (page 39)

1   a)   oxygen, water, glucose   (3)
    b)   Sun, light, chemical   (3)
    c)   green, chlorophyll   (2)
    d)   glucose   (1)
    e)   Roots   (1)
    f)   biomass   (1)
2   a)   iodine   (1)
    b)   from yellow/orange/brown to blue/black/blue-black   (1)
    c)   glucose that is not immediately used for respiration is converted to starch for storage   (1)
3   Removes carbon dioxide and adds oxygen   (1)
4   Plants are producers at the base of all first chains   (1)

Ask yourself (page 39)

No light, no photosynthesis, no energy production, no plant life, no producers in food chains, food chains collapse, no food, failure of life processes in animals, death of all living creatures.

8 Respiration
Exercise 8 (page 42)

1   a)   Respiration is a series of **chemical reactions** carried out within each living cell to **release energy** from chemical energy stores for all life processes.   (2)
    b)   in and around the mitochondria   (1)
2   a)   glucose + oxygen → carbon dioxide + water + energy   (6)
    b)   involves oxygen   (1)
3   a)   heart and lungs   (2)
    b)   Stored energy in glucose can be transferred to energy stores in the body without oxygen.   (1)
4   a)   yeast   (1)
    b)   alcohol/beer or bread   (1)

9 Reproduction in flowering plants
Exercise 9 (page 45)

1   a)   anthers   (1)
    b)   ovules   (1)
    c)   pollination   (1)
    d)   fertilisation   (1)
    e)   fruits   (1)
2   a)   one from: bright colours, pleasant smells   (1)
    b)   one from: parachutes (dandelions), wings (sycamore)   (1)
    c)   one from: juicy flesh (mango), barbed spines to attach to the fur of passing animals   (1)

10 Recreational drugs and human health
Exercise 10 (page 48)
1 A2, B4, C1, D3 (4)

2 **a)** Any two from: provision of safe drinking water, removal and safe disposal of sewage and rubbish, medical care such as immunisations and medicine. (2)

b) Any example of an infectious disease: e.g. measles, mumps, chicken pox, Ebola (1)

c) antibiotics (1)

3 512 (the number doubles every 20 minutes, 3 hours is 9 ×20 minutes, $2^9 = 512$) (2)

11 The interdependence of organisms in an ecosystem
Exercise 11A (page 49)
1 **a)** D (1)

b) D (1)

c) A (1)

d) A and C (1)

e) B (1)

Exercise 11B (page 52)
1 The number of herbivores will decrease quickly, fewer plants will be eaten and their population will grow. Reduction in fox numbers as fewer herbivores to eat. (3)

2 Introducing a new organism in to a food web; removing an organism from a food web; habitat destruction; overpopulation; waste from industry leaching into the environment; fossil fuel use polluting the environment. (3)

3 Reforestation; habitat protection; slow down extinctions to stop the fall in biodiversity; fossil fuel reduction; energy conservation. (3)

Ask yourself (page 52)
Answer depends on where you live and what issues your particular environment faces. Think how the environment may not be in the state it should be and think about what you would do to improve it. For example:

- Reduce pollution
- Tidy up and deal with refuse
- Plant trees and flowers
- Recycling opportunities

12 Classification of living things
Exercise 12A (page 55)
1 (5)

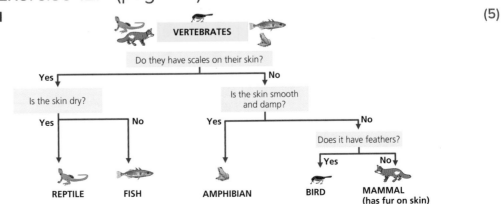

2 a) both invertebrates (1)
b) Insect (fly) has 3 main body parts, 6 legs, pair of antennae and usually 2 pairs of wings
Spider has 2 main body parts, 8 legs and no wings. (3)

3 a) and b) (8)

invertebrate no backbone	vertebrate backbone
octopus	cat
spider	frog
crab	fox
beetle	shark

4 They are both vertebrates, but a lizard has dry skin whereas a newt has smooth damp skin. (2)

Exercise 12B (page 58)

1 A3, B5, C2, D1, E6, F9, G4, H7, I8 (9)
2 Grass (a plant) is made up of many cells with cell walls, and it can carry out photosynthesis. (1)
Rabbits (vertebrate animals) are made up of cells that have no cell walls, they have a backbone and fur on their skin. (1)
Mushrooms (fungi) are made of cells that have cell walls but cannot carry out photosynthesis (so not green). (1)

13 Variations in living organisms

Exercise 13 (page 60)

1 a) Continuous variation because they can fall into very many groups. (1)
b) They result from **inherited genes** working with their environment. (1)
2 a) shoe size, blood group, eye colour (1)
b) Because they can be put into groups easily/they have a limited number of possible values. (1)
c) The others (continuous variations) are not easily put into discrete groups because there are many groups for each particular feature. (1)

Biology – Test yourself

1

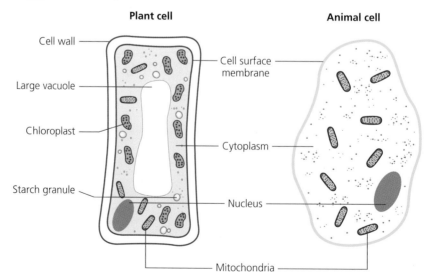

2   Tissue: cells of the same type and function combine together to make tissues, e.g. muscles, skin, blood
    Organ: a structure made from various tissues that combine to perform a specific function e.g. eye, leaf, heart.
    System: organs combine into a system to perform a specific job, e.g. reproductive system.

3   a) i)   carbohydrates (these are a chemical store of energy)
       ii) Any one from: foods containing sugar (glucose) – fruits, jams, sweets, soft drinks etc.; foods containing starch – potatoes, flour products, pasta etc.
    b) i)   protein
       ii) Any one from: meat, fish, milk, cheese, eggs, nuts

4   a) adolescence
    b) sperm cell, testis/testes
    c) egg cell/ovum, ovary/ovaries
    d) zygote
    e) Implantation
    f) pregnant

5   Any two from:
    • Exercise causes the heart to beat faster, which keeps it healthy.
    • Exercise develops muscles.
    • Exercise reduces the amount of stored fat in the body.

6   Plants make their own food by the process of photosynthesis. They use carbon dioxide from the air and water from the soil. The plant uses energy transferred by light from the Sun to the chlorophyll in the leaves to produce oxygen and sugars that are useful to all living organisms.

7   carbon dioxide and water

8   glucose (sugar) and oxygen

9   Some of the oxygen will be used by the plant for aerobic respiration. Oxygen not used will be released to the air through the underside of the leaf.

10 Glucose is needed by every living cells for respiration. The leaves make more glucose than they need, so glucose is transported to other parts of the plant for respiration.
    Energy stored in the food (glucose) that is not used immediately for respiration or growth is transferred to other chemical stores of energy such as starch.

11 a) in the soil (as soluble salts dissolved in water in the soil)
    b) roots

12 the differences between organisms of the same species

13 sorting organisms with similar characteristics (features) into groups

14 Both are vertebrates and belong to the class mammals. Both are warm-blooded, have skins that are partially covered with hair or fur. Both will have been born alive and fed on milk from the mother.

15 Jointed limbs in pairs, hard outer covering (exoskeleton) and bodies divided into segments.

16

Vertebrate	Invertebrate
emu	earthworm
shark	spider
whale	crab
frog	beetle
turtle	
fox	

17 a habitat together with all the organisms living in that habitat (all the living and non living parts)

18 the conditions within an ecosystem

19 a) rose

it is the only organism in the chain that can make its own food by photosynthesis

b) Any two from:

- aphids die out – killed by insecticide
- rose grows better – not being eaten by aphids
- robin and cat might move away – a supply of food has disappeared

20 a) a set of interconnected food chains

b) In a food web, consumers have more than one supply of food; a food chain, shows only one supply of food for each organism.

Chemistry

14 States of matter – the particle model

Exercise 14A (page 64)

1 a) Solid, liquid, gas (1)
   b) **Solid** Particles packed closely together, particles do not move around but vibrate about a fixed position, cannot be squashed into smaller volume. (1)
       Liquid Any three from: particles very close together, cannot be squashed, particles constantly move around each other, can flow, volume remains the same. (1)
       Gas Particles a long way from each other, will fill any container, can be squashed into a smaller space, move around in random directions. (1)
2 a) close together in fixed pattern do not move around but vibrate (2)
   b) packed closely together but are in a random arrangement, constantly moving around each other (2)
   c) a long way from each other, move around rapidly in all directions (2)
3 a) evaporation and condensation (2)
   b) i) evaporation (1)
       ii) condensation (1)
4 a) melting (1)
   b) freezing (1)

Exercise 14B (page 66)

1 A3, B4, C1, D2 (4)
2 a) sublimation (1)
   b) one of: carbon dioxide, iodine (1)
3 a) pressure (1)
   b) random (1)

15 Atoms and elements

Exercise 15 (page 69)

1 a) one (1)
   b) H (1)
   c) Cl (1)
   d) Periodic Table (1)
2 a) Any one from: iron, copper, magnesium, sodium, calcium or another metal (1)
   b) Any four from: solid at room temperature (except mercury), shiny, malleable (bendy), sonorous, usually very dense, good conductors of heat or/and electricity (4)
3 a) Any one from: hydrogen, carbon, nitrogen, oxygen, sulfur, chlorine, helium or another non-metal (1)
   b) Any four from: may be solid, liquid or gas at room temperature, dull, brittle, not sonorous, low density, poor conductors of heat or/and electricity, generally good insulators (4)

16 Compounds and molecules

Exercise 16 (page 72)

1 a) elements, chemical reaction, compound (3)
   b) molecule (1)
   c) products (1)

2 a) (9)

Substance	Chemical symbol or formula	Solid, liquid or gas (at room temperature)	Colour	Does it conduct electricity?	Any special property?
Oxygen	O	Gas	Colourless	Poor conductor	
Carbon	C	Solid	Black	No	
Helium	He	Gas	Colourless	No	
Iron	Fe	Solid	Grey-black	Yes	Magnetic
Copper	Cu	Solid	Reddish brown	Yes	Not magnetic
Water	H_2O	Liquid	Slight blue tint	No	
Iron sulfide	FeS	Solid	Black	No	
Methane	CH_4	Gas	Colourless	No	
Sodium chloride	NaCl	Solid	Colourless	No	

   b) water, iron sulfide, methane, sodium chloride (1)

17 Pure substances

Exercise 17 (page 74)

1 a) pure (1)
   b) 100°C (1)
   c) Evaporation (1)
   d) increase (1)
2 It expands when it freezes; ice floats on liquid water. (2)
3 Heat makes particles in a liquid move around more quickly. Eventually enough energy will have been transferred to allow particles to escape from the liquid and the liquid becomes a gas (1)
4 Check the liquid boils at 100°C
   Evaporate the water to dryness – if there is no residue, it is pure water. (2)

18 Mixtures including solutions

Exercise 18 (page 78)

1 a) a compound (1)
   b) seawater (1)
   c) solute (1)
   d) solvent (1)
   e) the same as (1)
   f) randomly (1)

2   Oxygen is used in aerobic respiration, which is a series of chemical
     reactions that release energy. Energy is needed for life processes
     to continue.   (2)
3   a)   needed for aerobic respiration   (1)
     b)   needed for the process of photosynthesis   (1)

19 Separating mixtures

Exercise 19A (page 80)

1   a)   soluble: substances that dissolve in a liquid   (2)
          insoluble: substances that do not dissolve in a liquid   (2)
     b)   The sulfur does not dissolve in the water; the copper sulfate does,
          making the water turn blue.   (2)
     c)   i)   sulfur   (1)
          ii)   (hydrated) copper sulfate solution   (1)
2   a)   solids, different   (2)
     b)   insoluble, liquids   (2)
     c)   filtrate   (1)
     d)   decanting   (1)
3   a)   solvent   (1)
     b)   evaporated, solute   (2)

Exercise 19B (page 82)

1   (12)

Elements	Compounds	Mixtures
magnesium	carbon dioxide	air
carbon	sodium chloride	seawater
iron filings	distilled water	crude oil
oxygen	iron sulfide	dilute sulfuric acid

2   a)   (6)

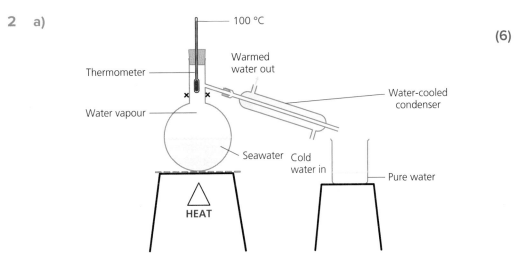

     b)   Check the liquid boils at 100°C
          Evaporate the water to dryness – if there is no residue, it is pure
          water.   (2)
3   a)   evaporation   (1)
     b)   filtering   (1)

20 Separating and identifying mixtures

Exercise 20 (page 84)

1   method named as chromatography (1)
    Any three points from the following: (3)
    - put a dot of the black ink on the bottom of a piece of chromatography paper
    - dip the bottom edge of the paper in a beaker of water
    - watch the water dissolving the different substances in the ink
    - the most soluble will travel the furthest up the paper
    - the least soluble travels the shortest distance
2   Substance A is more soluble than substance B. (1)
3   compounds in drugs, blood samples (1)

21 Combustion – chemical reactions

Exercise 21 (page 89)

1   a)   a chemical change caused by burning (1)
    b)   fuel and oxygen (2)
2   Oxygen combines with another element to form an oxide. This is called oxidation. Combustion is an example of this. (2)
3   During a reaction (chemical or physical) no particles are lost or gained, so the total mass of the substances involved does not change. (1)
4   **sulfur + oxygen → sulfur dioxide** (1)
5   CO_2 displaces the oxygen the fire needs in order to burn (1)

Ask yourself (page 89)

Effects on ecosystems: smaller areas of habitat suitable for organisms living in the affected biomes (forest, grassland/savannah, peatland); food chains and webs may collapse as plants at their base are burned; plants or/and animals may become extinct.

Effects on the atmosphere: increased amounts of CO_2 in the atmosphere leading to increased greenhouse effect and greater climate change; smoke in the atmosphere may cause health problems for those living nearby/downwind.

22 Fuels and production of carbon dioxide

Exercise 22 (page 93)

1   a)   a compound that contains only hydrogen and carbon (1)
    b)   water and carbon dioxide (2)
    c)   **methane + oxygen → carbon dioxide + water** (4)
2   a)   A fuel that is formed from the remains of dead animals and plants that lived millions of years ago. (1)
    b)   oil, natural gas (2)
    c)   i)   Two from: nitrogen dioxide, carbon dioxide, sulfur dioxide. (2)
        ii) They form airborne acids (nitric acid, carbonic acid and sulfuric acid respectively). (2)
3   A fuel is a store of energy.
    When a fuel is burned, energy is transferred from one store to another or from this store of energy to the surroundings. (2)

4 a) Any four points from: (4)
   - The Earth is warmed by radiation from the Sun. When the radiation reaches the Earth, it is reflected back into the atmosphere.
   - Layers of greenhouse gases stop some of this reflected heat from escaping and it is reflected back to the Earth's surface.
   - The surface of the Earth is made warmer (global warming).
   - Carbon dioxide is a greenhouse gas.
   - Carbon dioxide is added to the atmosphere when fossil fuels are burned.

   b) Any four from: increasingly strong winds, more deserts, increased flooding, melting ice caps, heavier rain, pests moving to new places (4)

   c) Any two from: reduce carbon emissions, stop burning fossil fuels, reduce deforestation, plant more forests (2)

   d) Burning forests releases carbon dioxide into the atmosphere. Clearing forests reduces the number of plants taking carbon dioxide from the atmosphere (for photosynthesis). (2)

Ask yourself (page 93)

Reduce the burning of fossil fuels, particularly coal. Conserve energy in your homes, work places and in industry. Switch from fossil-fuel powered cars to electric vehicles.

23 Oxidation of metals
Exercise 23 (page 97)

1 a reaction with oxygen (1)
2 a) yes, magnesium oxide (1)
   b) yes, magnesium hydroxide and hydrogen (1)
   c) no reaction (1)
   d) yes (slowly), zinc sulfide and hydrogen (1)
3 Because it does not react with water or acids (and is a good conductor of heat). (1)
4 a) Iron reacts easily with water and oxygen so it is a common reaction. It costs money to replace items that have rusted. It is expensive to put measures in place to stop rusting. (2)
   b) iron, oxygen and water (3)
   c) Any three of the following: (3)
      Cover iron in:
      - a layer of grease or oil
      - paint
      - plastic
      - zinc
      - tin

24 Thermal decomposition reactions
Exercise 24 (page 99)

1 a) decomposition (1)
   b) thermal decomposition (1)
2 Bubble the gas through limewater to test for carbon dioxide. If it goes milky the gas is carbon dioxide. (1)
3 Check the liquid boils at 100°C
   Evaporate the water to dryness – if there is no residue, it is pure water. (2)

25 Acids and alkalis

Exercise 25 (page 102)

1 a) (acid) red   (1)
   b) (alkaline) blue   (1)
   c) (alkaline) blue   (1)
   d) (neutral) will remain purple   (1)
   e) should remain purple, but depends on whether the water is acidic
       or not   (1)

2 **pH scale** (a scale of numbers ranging from 1 to 14).

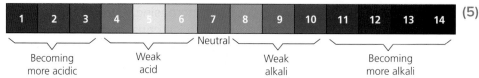
                                                                                           (5)

3 a) a substance that can neutralise acids   (1)
   b) Any two suitable answers: e.g. sodium hydroxide, calcium hydroxide,
       indigestion tablets, baking powder, toothpaste   (2)
   c) a base that can dissolve in water   (1)
   d) So that it can neutralise the acid in our mouths. Acid can cause tooth
       decay so an alkaline toothpaste can help prevent this happening.   (1)

4 a) **acid + base → salt + water**   (1)
   b) Any suitable example: e.g.
       hydrochloric acid + sodium hydroxide → sodium chloride + water   (1)
   c) an acidic solution such as vinegar   (1)
   d) an alkali solution such as baking powder   (1)

26 Neutralisation reactions

Exercise 26 (page 105)

1 a) salt   (1)
   b) salt   (1)
   c) salt, carbon dioxide   (2)
   d) salt, hydrogen   (2)
   e) lowers   (1)
   f) raises/increases   (1)
   g) it increases   (1)

2 a) i)   **base + acid → salt + water**   (2)
       ii) any suitable example: e.g.
            magnesium oxide + hydrochloric acid → magnesium chloride + water
            zinc oxide + hydrochloric acid → zinc chloride + water   (2)
   b) i)   **metal + acid → salt + hydrogen**   (2)
       ii) any suitable example: e.g.
            zinc + hydrochloric acid → zinc chloride + hydrogen   (2)
   c) i)   **metal oxide + acid → salt + water**   (2)
       ii) any suitable example: e.g.
            zinc oxide + hydrochloric acid → zinc chloride + water   (2)
   d) i)   **metal carbonate + acid → salt + water + carbon dioxide**   (2)
       ii) any suitable example: e.g.
   copper carbonate + sulfuric acid → copper sulfate + water + carbon dioxide   (2)

3 Bubble gas through limewater. If carbon dioxide is present, the
   limewater will turn milky.   (2)

4 Place a lighted splint in a test tube containing the gas. If the gas is hydrogen, there will be a squeaky pop. (2)

Ask yourself (page 105)

As a building material it is abundant and versatile. It is easy to carve and cut. It looks good and is strong.

Chemistry – Test yourself

1 a)

	Solids	Liquids	Gases
Arrangement of particles	Particles packed closely together	Particles are very close together	Particles a long way from each other so move around rapidly
Do they flow easily?	No	Yes, because particles constantly moving round each other	Yes
Can they be compressed?	No	No	Yes
Can they change their shape?	No	Yes	Yes

b) i) Energy used to raise the temperature of the substance and bonds between particles in the ice begin to break. Ice turns from a solid to a liquid, water.

ii) Further heating of the water causes further separation of the particles and the liquid evaporates to form a gas in which the particles are a long way from each other.

c) when a substance changes straight from solids to gases

d) i) Particles in a gas move around quickly and randomly. As they bump into the walls of the container, they create a force so causing pressure.

ii) As the gas is heated, the particles move around faster, so they bump into the walls with more force more often and the pressure in the football will increase.

2 a) i) molecules or compounds

ii) element

iii) one

iv) identical atoms

v) metals, non-metals

b) If the lamp lights when the switch is turned on the substance is a conductor.

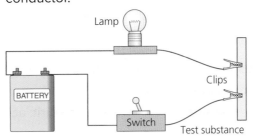

3 a) i) compound

ii) chemical bonds

b)   i)   any example of a compound with name and formula
           e.g. carbon dioxide, CO_2; water, H_2O

     ii)   any example:
           CO_2 carbon (C) and oxygen (O)
           H_2O hydrogen (H) and oxygen (O)

     iii)   No, different properties

4   a)   i)   mixture

        ii)   physical

        iii)   chemical

        iv)   the same as

   b)   stir the mixture with a magnet

5   a)   when you want to recover a solvent from a solution

   b)   when the different substances have different boiling points

   c)   Any suitable example: e.g. producing pure water from seawater or pure
       water from ink solution

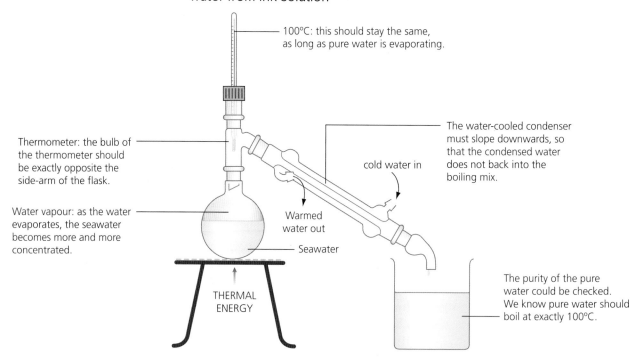

100°C: this should stay the same, as long as pure water is evaporating.

Thermometer: the bulb of the thermometer should be exactly opposite the side-arm of the flask.

The water-cooled condenser must slope downwards, so that the condensed water does not back into the boiling mix.

cold water in

Water vapour: as the water evaporates, the seawater becomes more and more concentrated.

Warmed water out

Seawater

THERMAL ENERGY

The purity of the pure water could be checked. We know pure water should boil at exactly 100°C.

6   a)   i)   combustion

        ii)   heat

       iii)   oxygen, oxide, oxidation

   b)   possible examples:

     •   burning of magnesium in oxygen
        magnesium + oxygen → magnesium oxide

     •   burning of carbon in oxygen
        carbon + oxygen → carbon dioxide

7   a)   When the coal is burned the sulfur combines with the oxygen and sufur
       oxides are released into the atmosphere.

     sulfur + oxygen → sulfur dioxide

   b)   i)   **sulfur dioxide + water → sulfuric acid**

        ii)   acid rain

   c)   Any two from:

     •   Damage to aquatic environments (lakes, rivers etc.) that affects fish
       and organisms in these habitats.

     •   Damage to forests as the acid rain damages the soil.

     •   Damage to human-made structures particularly those made of
       limestone or marble.

8 a) i) **metal + water → metal hydroxide + hydrogen**

ii) **magnesium + water → magnesium hydroxide + hydrogen**

b) i) **metal + acid → salt + hydrogen**

ii) Any suitable example such as:
magnesium + sulfuric acid → magnesium sulfate + hydrogen

9 a) **copper carbonate → copper oxide + carbon dioxide**

b)

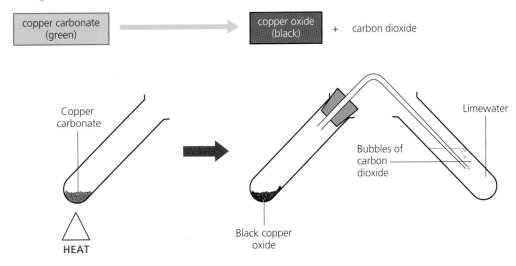

c) Bubble the gas through limewater. If the limewater goes milky, the gas is carbon dioxide.

10 a) changes appearance, so indicating the presence of particular substances

b) Any two from:
- limewater: to test for the presence of carbon dioxide
- anhydrous copper sulfate: to test for the presence of water
- litmus: to test for presence of acids and alkalis
- Universal Indicator: to test for the presence and strength of acids and alkalis

c) i) stays red
ii) turns blue

d) i) **acid + base → salt + water**
ii) any suitable example, such as:
hydrochloric acid + sodium hydroxide → sodium chloride + water

e) i) **metal + acid → salt + hydrogen**
ii) Any suitable example, such as:
zinc + sulfuric acid → zinc sulfate + hydrogen

f) i) **metal oxide + acid → salt + water**
ii) Any suitable example such as:
copper oxide + sulfuric acid → copper sulfate + water

iron oxide + sulfuric acid → iron sulfate + water

g) i) **metal carbonate + acid → salt + water + carbon dioxide**
ii) Any suitable example such as:
copper carbonate + sulfuric acid → copper sulfate + water + carbon dioxide

Physics

27 Energy resources

Exercise 27 (page 111)

1 a) fuels formed from fossilised remains of animals and plants that
       lived many millions of years ago   (2)
    b) Any two from: coal, natural gas, oil   (2)
    c) once burned they cannot be used again
       we are using them up faster than they can be replaced (if they can
       be replaced at all) so they will eventually run out   (2)
    d) No. Fossil fuels provide raw materials for the manufacture of
       plastics, medicines, cosmetics and synthetic fibres.   (2)
    e) advantages: often easy to extract, can be used at any time   (2)
       disadvantages: supplies running out, cannot be replaced, very
       polluting to the environment because they release greenhouse gases   (2)
2 a) energy resources that do not get smaller as they are used   (2)
    b) any two from: solar power, biomass, wind power, geothermal power,
       hydroelectric power, tidal power, wave power   (2)
    c) advantages: not used up, little or no atmospheric pollution,
       becoming cheaper to provide   (2)
       disadvantages: unreliable, only available in certain places, can be
       expensive to use, can be bad for the environment   (2)

28 Energy transfers and conservation of energy

Exercise 28 (page 114)

1 A5, B4, C1, D3, E2, F6   (6)
2 In an energy chain, the total amount of energy stays the same even
   though it is being transferred from one store to another.
   **total amount of energy stored = the total amount
   at the start   of energy stored at the end**   (2)
3 a) kinetic store of energy transferred to a thermal store of energy in
       the surroundings (energy may be lost to sound)   (2)
    b) gravitational store of energy to kinetic store of energy
       (energy may be lost to heat and sound due to friction)   (2)
    c) chemical store of energy transferred to a thermal store of energy
       in the surroundings (energy may be lost as heat)   (2)

29 Speed and movement

Exercise 29 (page 117)

1 a) $\text{speed} = \dfrac{\text{distance}}{\text{time}}$ (1)

    b)   (4)

Race distance, in m	Time taken, in s	Speed, in m/s
100	9.58	10.4
800	101	7.9
5 000	755	6.6
10 000	1 571	6.4

2 a) 210 m   (1)
    b) 250 s   (1)

3   21 km/h   (1)
     (relative speed = fastest speed – slowest speed; 86 – 65 = 21)
4 a) different objects, ruler/measure (up to 4 m in length), a stopwatch
        (or computer and light gates), a pen and paper (or a computer) to
        record results   (4)
   b) • Measure a height of 4 m and make a mark at this point.
       • Take the first object and drop from the marked height of 4 m.
       • Use a stopwatch (or a computer and light gates) to time how long it
         takes to reach the floor.
       • Start the stopwatch as the object is released and stop it when it
         reaches the floor.
       • Repeat this process several times and calculate a mean
         (average) time.
       • Record your results.
       • Repeat this process for other objects.   (4)

   c) $$\textbf{mean speed} = \frac{\textbf{distance (4 m)}}{\textbf{mean time taken for object to reach ground}}$$ (2)

   d) Only change one variable and keep all other variables constant
        throughout.   (1)

30 Measuring force

Exercise 30A (page 119)

1   Make a moving object: (3)
    • change speed
    • change direction
    • change shape
2 a) push   (1)
   b) newtons   (1)
   c) Gravity   (1)
   d) Friction   (1)
3   The gravitational force pulling the boat downwards is balanced by the
    upward support force (upthrust) exerted by the water on the boat.   (2)

Exercise 30B (page 122)

1 a) 30 N   (2)
   b) 300 N   (2)
   c) 1800 N   (2)
   d) 5 N (0.5 kg × 10)   (2)
   e) 3.2 N (0.32 kg × 10)   (2)

2 a) 80 kg   (1)
   b) 800 N   (2)
   c) 128 N   (2)

Ask yourself (page 122)

In your experiments you should have found that the rougher the surface, the
shorter the distance travelled by the toy car. Friction is always more on rough
surfaces.

31 Effects of forces

Exercise 31 (page 125)

1 a) increase, go faster (2)
   b) decrease (slow down) (1)
   c) constant speed in the same direction (or remain at rest) (2)

2 (2)

FORCE PRODUCED BY ENGINE

AIR RESISTANCE

FRICTION

3 a) gravity (1)
   b) i)   air resistance (1)
       ii) the gravitational force pulling the spaceship downwards is balanced by the upthrust of the water on the spaceship (2)

4 Equipment: paper, scissors, string, sticky tape, model parachutist (modelling clay), stopwatch, metre rule, pen and paper (or computer) to record results.
   Method:
   - Use scissors to cut out parachutes with different areas.
   - Attach model parachutist to the first parachute.
   - Drop the parachute from a measured height.
   - Record the time taken to reach the ground.
   - Record your results.
   - Repeat this drop several times so that you can calculate an average speed.
   - Repeat this experiment for all the parachutes. (6)

Ask yourself (page 125)

The racing cars are designed to minimise the effect of air resistance. Air resistance is the friction between the air and a moving object. The Formula 1 racing teams want their cars to travel as fast as possible. Designers spend huge amounts of time and money examining how the air flows over model car bodies to make cars that are streamlined and more aerodynamic. They also aim to increase the downward pressure on the tyres, so friction between the tyres and the road is increased and there is more grip. The F1 wing is also designed to increase downward force (it operates in the opposite way to an aeroplane wing). The cars also have a flap on the rear wing (DRS), which is activated to produce an even more streamlined shape. There is much to investigate!

32 Pressure

Exercise 32 (page 127)

1 a) $\text{pressure} = \dfrac{\text{force}}{\text{area}}$ (1)
   b) force = pressure × area (1)
   c) newtons per square metre (N/m^2) or newtons per square centimetre (N/cm^2) (1)

2 a) The force is spread over a large area so decreasing the pressure.
       This enables the tractor to travel over muddy ground without sinking. (1)
   b) The force is exerted on a small area which increases the pressure,
       enabling the blade to cut through the earth more easily. (1)
   c) $P = \dfrac{F}{A}$
       $= \dfrac{1200\,\text{N}}{2\,\text{cm}^2}$
       $= 600\,\text{N/cm}^2$ (2)

33 Density

Exercise 33 (page 131)

1 mass and volume (1)

2 $\textbf{density} = \dfrac{\textbf{mass}}{\textbf{volume}}$ (1)

3 a) $75\,\text{g/cm}^3$ (1)
   b) $11\,\text{g/cm}^3$ (1)
   c) $2.7\,\text{g/cm}^3$ (1)
   d) $2.7\,\text{g/cm}^3$ (1)
4 C and D; because they have the same densities (2)
5 glass (1)
   Use $\textbf{volume} = \dfrac{\textbf{mass}}{\textbf{density}}$

   For marble: $\textbf{volume} = \dfrac{3000\,\textbf{g}}{3.2\,\textbf{g/cm}^3} = 937.5\,\textbf{cm}^3$ (1)

   For glass: $\textbf{volume} = \dfrac{3000\,\textbf{g}}{2.8\,\textbf{g/cm}^3} = 1071.4\,\textbf{cm}^3$ (1)

6 volume of room is $10\,\text{m} \times 6\,\text{m} \times 3\,\text{m} = 180\,\text{m}^3$ (1)
   $\textbf{mass} = \textbf{density} \times \textbf{volume}$
   so mass $= 1.3\,\text{kg/m}^3 \times 180\,\text{m}^3 = 234\,\text{kg}$ (1)

34 Sound

Exercise 34 (page 135)

1 a) vibrates (1)
   b) sound waves (1)
   c) particle (1)

2 a) The particles are closer together, can transmit the vibrations
       faster and have stronger bonds (1)
   b) There are no particles in a vacuum, so the vibrations cannot be
       passed on. (1)

3 a) i)   frequency (1)
       ii) amplitude/size (1)
   b) The high, loud note has a higher frequency and larger amplitude
       than a low, quiet one. (2)

4 A dog whistle produces a note with a very high frequency. Dogs can
   hear sounds with such high frequencies, but humans cannot. (2)

35 Light

Exercise 35A (page 138)

1   It is reflecting light from a luminous source into our eyes. (2)

2 a) transparent (1)
     b) opaque (1)
     c) translucent (1)

3 a) incident rays, angle of incidence. (2)
     b) reflected rays, angle of reflection. (2)

4 a) it is reflected (1)
     b) angle of incidence = angle of reflection (1)

Exercise 35B (page 142)

1   Any three from: travel through a vacuum; travel in straight lines; travel very fast; travel faster than sound; will not travel through opaque materials; can be absorbed/reflected/scattered. (3)

2 a) the bending of a light ray (1)
     b) at a boundary between different mediums (1)
     c) light travels at different speeds in different mediums (1)
     d) Light rays reflected from the coin pass from water to air on the way to our eyes. They move from a dense medium (the water) to a less dense medium (air) and bend away from the normal so the apparent position of the coin will shift.

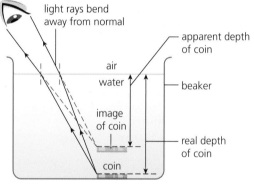

(3)

3 a) (4)

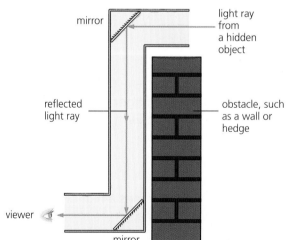

     b) a periscope (1)
     c) Any suitable use such as: submarine looking above the water line; a shopkeeper keeping a watch on his stock in another aisle; observing things form a hiding position. (1)

4   a)   (2)

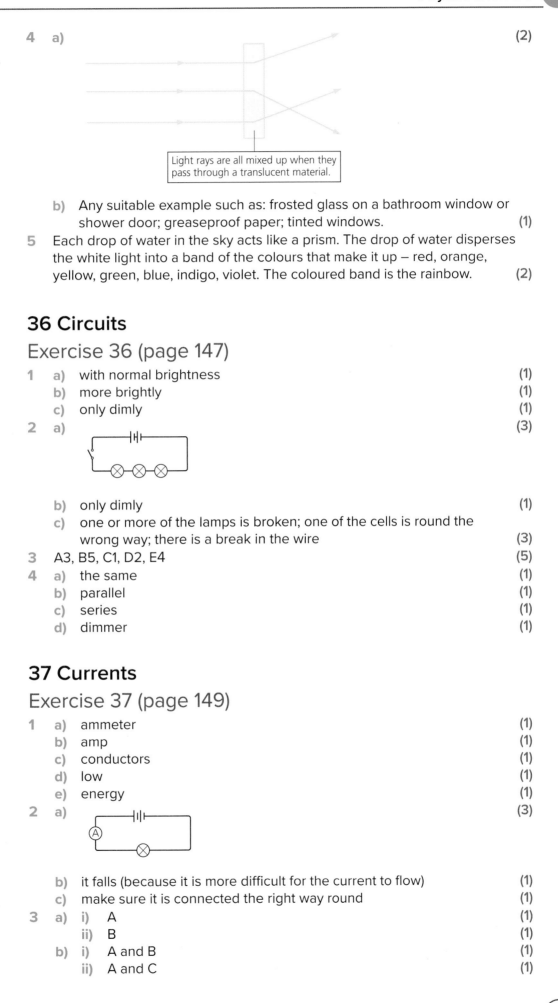

Light rays are all mixed up when they pass through a translucent material.

b)   Any suitable example such as: frosted glass on a bathroom window or shower door; greaseproof paper; tinted windows.   (1)

5   Each drop of water in the sky acts like a prism. The drop of water disperses the white light into a band of the colours that make it up – red, orange, yellow, green, blue, indigo, violet. The coloured band is the rainbow.   (2)

36 Circuits
Exercise 36 (page 147)

1   a)   with normal brightness   (1)
     b)   more brightly   (1)
     c)   only dimly   (1)
2   a)   (3)

b)   only dimly   (1)
c)   one or more of the lamps is broken; one of the cells is round the wrong way; there is a break in the wire   (3)
3   A3, B5, C1, D2, E4   (5)
4   a)   the same   (1)
     b)   parallel   (1)
     c)   series   (1)
     d)   dimmer   (1)

37 Currents
Exercise 37 (page 149)

1   a)   ammeter   (1)
     b)   amp   (1)
     c)   conductors   (1)
     d)   low   (1)
     e)   energy   (1)
2   a)   (3)

b)   it falls (because it is more difficult for the current to flow)   (1)
c)   make sure it is connected the right way round   (1)
3   a)   i)   A   (1)
          ii)   B   (1)
     b)   i)   A and B   (1)
          ii)   A and C   (1)

c) i) 1.2 A   (1)
    ii) 2.2 A   (1)

38 Magnetic fields and electromagnets

Exercise 38A (page 151)

1 a) magnetic   (1)
   b) north-seeking pole (of the magnet)   (1)
   c) Unlike; like   (1)

Exercise 38B (page 155)

1   (2)

2 a) Wrap the insulated wire around the iron nail. Attach the ends of the
       wire (with insulation removed) to the cell. When the current is
       switched on, the iron nail will become a magnet.   (3)
   b) Get some unmagnetised steel paperclips/pins (or similar) and see
       whether they are attracted to the iron nail or become attached to it.
       If they do, it shows the nail has become a magnet.   (2)

   c) add more turns to the coil; increase the current   (2)

39 Space

Exercise 39A (page 157)

1 The Earth spins on its axis, completing one turn every 24 hours. During this
   time, half of the Earth faces the Sun and experiences daytime, while the other
   half is in darkness – night-time.   (2)
2 the spinning of the Earth   (1)
3 At midday the Sun is high in the sky and shadows are shortened.
   In the evening the Sun is low in the sky and shadows are long.   (2)
4 27 days; the time it takes for the Moon to orbit the Earth   (2)
5 The Moon is not a light source so it is hard to see during daytime.
   We see it best at night because the light from the Sun is reflected from it. (2)

Exercise 39B (page 162)

1 The Earth is tilted on its axis. This means that at any one time, part of the
   Earth's surface is closer to the Sun than at other times. The Earth will tilt
   towards the Sun in summer and away from the Sun in winter.   (2)
2 a) Earth, Sun   (2)
   b) Moon, Earth   (2)
3 the distance light travels in 1 year   (1)
4 In space there is less light pollution and no atmospheric turbulence.
   The Hubble telescope can detect infrared and ultraviolet light waves,
   which are absorbed by the Earth's atmosphere before they are detected
   by land-based telescopes.   (2)

Ask yourself (page 162)

We depend on space technology. If satellites were to be wiped out (as a result of a solar storm or sabotage, for example) we might experience the following:

- No television and radio.
- No GPS (Global Positioning System) for navigation. Computer systems also depend on this so parts of the Internet may fail.
- No communication satellites would lead to problems with international telephone calls and some Internet connections; without these and secure satellite phones, world leaders wouldn't be able to communicate.
- Power systems fail leading to power cuts; water treatment plants fail leading to disease.
- Failure of transport routing systems from air traffic control to traffic lights.
- No weather forecasting.

The list is endless. We rely more on satellites than we realise.

Physics – Test yourself

1. a) a ruler; cm
   b) balance scales; g
   c) a ruler (measure width and height and multiply to find the volume), cm^3
   d) a measuring cylinder (pour remaining water into the cylinder and read off the volume), marked in cm^3
   e) a measuring cylinder with some water in it (measure the volume of water that the keys displace), cm^3

2. Volume of small box: 2 cm × 3 cm × 0.5 cm = 3 cm^3
   Volume of large box: 60 cm × 30 cm × 15 cm = 27 000 cm^3
   Number of drawing pin boxes that can fit into the box: 27 000 cm^3 ÷ 3 cm^3 = 9000

3. Volume of large box: 30 cm × 50 cm × 20 cm = 30 000 cm^3
   mass = volume × density
   mass of water = 30 000 cm^3 × 1 g/cm^3 = 30 000 g (30 kg)

4.
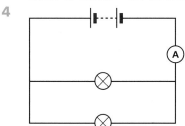

5. weight (N) = mass (kg) × gravitational force (N/kg)

6. a) 7 cm
   b) 7.5 cm

7. a) incident ray
   b) It is reflected.
   c) The angle between the reflected ray and the normal is the same as the angle between the incident ray and the normal.

8

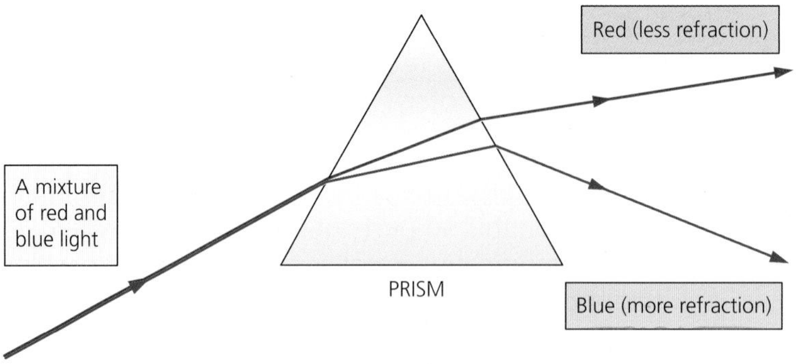

Red (less refraction)

A mixture of red and blue light

PRISM

Blue (more refraction)

9   For the Northern hemisphere:

Season	Height of Sun	Length of shadow	Length of day
Summer	High	Short	Long
Winter	Low	Long	Short

10   planet   star   solar system   galaxy   universe
11   the measure of the ability to do work
12   Any five from:
   - kinetic store of energy
   - gravitational potential store of energy
   - elastic store of energy
   - internal (heat/thermal) store of energy
   - nuclear store of energy
   - chemical store of energy
13   a)   make a moving object change speed; make a moving object change direction; change the shape of an object
     b)   Any four from: magnetic force, gravitational force, frictional force (contact and air resistance), reaction (support) force; upthrust
     c)   newtons (N)
     d)   i)   If an unbalanced force acts on a moving object in the opposite direction from the movement, then the object will decrease in speed (i.e. slow down or stop).
              If an unbalanced force acts on a moving object in the same direction as the movement, then the object will increase in speed (i.e. start moving or go faster).
              An unbalanced force may also cause an object to change direction.
          ii)   Any appropriate example: sinking, falling objects, pedalling a bicycle ...

Glossary

Biology

aerobic respiration The form of respiration involving oxygen

anaerobic respiration When no oxygen is involved in respiration

antagonistic pairs A pair of muscles that have opposite actions enabling movement to take place

anther Part of the stamen, the male part of a flower

arthropods An arthropod is an invertebrate animal with an exoskeleton, a segmented body and paired and jointed limbs

bacteria Single-celled organism, some are harmful, others are not

brain An organ found in the head that controls many of the life processes in animals

carbohydrate A food substance, including starches and sugars; they are a chemical store of energy

carnivores Animals that eat other animals

carpel The female part of a flower – made up of stigma, style and ovary

cell surface membrane The part of a cell surrounding the cytoplasm

chlorophyll The green pigment in plant cells that can absorb light from the Sun for photosynthesis

consumers Organisms in a food chain that eat other organisms

cytoplasm All living material of a cell, other than the nucleus

fat A food substance providing a chemical store of energy

fertilisation The joining together of the sperm and egg in animals

fetus Unborn offspring of a mammal

fibre A substance that comes from plants that provides bulk to our food to enable it to travel through the digestive system more efficiently

filament Part of the stamen, the male part of a flower

flower Part of the plant containing the reproductive organs

food chain A list showing the flow of food energy between living organisms

food web A set of different food chains that overlap and link up with one another

fungi One of the five kingdoms of living organisms that has similar cells to plants but is unable to carry out photosynthesis

gamete Special sex cells – the male (sperm/pollen) or female (ovum)

genes Sections of the chromosomes that control the characteristics of living organisms

glucose The sugar obtained from digested food, which reacts with oxygen in respiration

growth A life process where an organism increases in size and/or number of cells

heart An organ that pumps blood through all parts of an animal's body

herbivores Animals that eat plants

intestines A long tube that runs from the stomach to the anus and breaks the food down, and where useful substances are passed into the blood

invertebrates Animals that have no backbone

lungs Organs that allow oxygen to enter the body and carbon dioxide to leave it

mineral salts Substances that usually combine with another food to form different parts of the body, such as teeth, bones (from calcium) and red blood cells (from iron)

nucleus contains information that the cell uses to make proteins that the cell uses to grow and stay alive

nutrients The food substances required to carry out the processes that are essential for life

nutrition The life process that provides a living organism with its food

obesity An extremely heavy body weight condition that often causes illness

omnivore An animal that eats both plants and animals

ovary In a mammal the ovary contains the follicles that develop into eggs

ovum The female gamete

oxygen The gas required by all living organisms in order to burn up food for energy

photosynthesis The process by which green plants use sunlight to make nutrients from carbon dioxide and water

pollination the transfer of pollen from the anther (male part) to the stigma (female part) of the plant, enabling fertilisation and reproduction

predator An animal that hunts, captures and often heats other animals

prey An animal hunted and captured by other animals

protein A food substance used in the growth and repair of cells

reproduction The life process that produces new individuals of the same species

respiration The release of energy from food molecules

roots The part of a plant that absorbs water and minerals from the soil; they also anchor the plant firmly in the soil

single-celled organism One of the five kingdoms of living organisms where the organism has a single cell with a nucleus

sperm A male gamete

stamen The male part of a flower made up of the filament and anther

starch A type of carbohydrate that is stored in the muscles and liver in humans and is the food store in many plant tissues

stem The part of the plant that supports the leaves, holding them up towards the light

Stigma The part of the carpel where the pollen grain lands; the female part of a flower

style A part of the carpel, the female part of a flower

vertebrate An animal with a backbone (as part of a bony skeleton)

viruses Microbes that invade living cells in order to reproduce

vitamins Substances needed in very small amounts to enable the body to use other nutrients more efficiently: for example vitamin C, which is crucial in avoiding bleeding gums and loose teeth

water The liquid formed from a combination of hydrogen and oxygen and required by all living organisms in order to survive

zygote The fertilised egg produced when gametes are joined together during fertilisation

Chemistry

atom   The basic unit that makes up an element

boiling   A physical change in which heat changes a liquid into a gas

boiling point   The temperature at which a liquid changes to a gas or a gas changes to a liquid

chemical change   A chemical change occurs when a chemical reaction causes a substance or substances to change into a new substance. Atoms are rearranged but the number of atoms does not change

compound   A chemical substance made of different elements linked to one another

condensation   A change of state from gas to a liquid, speeded up by cooling

conductors   Materials that are good at transferring heat or/and electricity (a metal wire is a conductor because it allows electricity and heat to pass through it)

decanting   A way of separating a solid from a liquid by letting the solid settle and then pouring the liquid into another container

decomposition   A chemical reaction in which one substance is broken down into several products

density   The amount of mass in a specified volume – usually the mass of 1 cm^3

dissolving   A process that spreads out particles of a solid through a liquid to produce a solution

distillation   The process of separation using evaporation and condensing that depends on substances in a solution or mixture having different boiling points

element   A substance that is made up of one type of atom; can not be broken into anything simpler

evaporating   A change of state in which a liquid changes into a gas, speeded up by heating

filtrate   The liquid that has passed through a filter

filtration   A process of separation that uses a filter (like a sieve) to separate an insoluble solid from a liquid

freezing   The physical change of a liquid into a solid as the temperature falls; this is also known as solidifying

insoluble   Something that will not dissolve in a liquid (sand is insoluble in water)

insulators   An insulator is a material such as polystyrene that does not allow heat/ electricity to pass through it

litmus paper   Test papers used to detect the presence of acids or alkalis

magnet   A substance that can attract a metal such as iron

melting   A physical process in which heat changes a solid to a liquid

melting point   The temperature at which a solid changes to a liquid or a liquid changes to a solid

metals   Elements that usually conduct heat and electricity

molecule   Two or more atoms joined together

neutralisation   A chemical reaction between acids and alkalis, producing a neutral solution

oxidation   A chemical reaction where elements combine with oxygen to form compounds called oxides

Periodic Table   A chart that arranges all the elements in order of their atomic number and in groups and periods according to their properties

pH scale   A scale of numbers, from 0 to 14, that gives a measurement of acidity or alkalinity

physical change   A physical change such as changes in state and dissolving are reversible. No new substances are formed

residue   The solid material left on a filter when a mixture is poured through it

rusting   A process in which air (oxygen) and water cause a chemical change to iron

saturated solution   A solution that has the maximum amount of solute dissolved in it at a particular temperature

sieving   The process of using a mesh to separate a mixture of solid particles of different sizes

soluble   Able to dissolve in a liquid, e.g. salt is soluble in water

solute   A substance that can dissolve in a liquid (the solvent) to form a solution

solution   The mixture formed when a solute dissolves in a solvent

solvent   A liquid that can dissolve a solute to form a solution

sublimation   The change of state from solid to gas or gas to solid, missing out the liquid state

Universal Indicator   a sensitive indicator of pH, which is red/orange under acidic conditions, purple/blue under alkaline conditions and green when neutral

water cycle   The change of water between solid, liquid and gas that circulates water around the planet Earth

Physics

air resistance Friction between the air and a moving object (sometimes called drag)

attract To pull towards each other, e.g. unlike poles of a magnet attract each other

cell A chemical store of energy; cells may be connected in series to make up a battery

circuit The complete route from the positive terminal to the negative terminal of a power source

electricity An energy pathway by which energy is transferred from one store to another store

force Something causing i) a stationary object to move and/or change shape; ii) change in the speed or the direction of movement of a moving object. This can be a push, pull, support (reaction) or upthrust

fossil fuels Coal, oil and gas: fuel that was made millions of years ago from the bodies of dead animals and plants

friction A force that opposes motion between two objects when they rub together

gravitational potential energy The energy that an object has because of its position and its mass; the vertical height from which the object falls and the force of gravity

gravity A force of attraction between any two objects

joule A unit of energy (energy transferred when a force of 1 newton is applied over a distance of 1 metre)

kinetic store of energy Energy transferred to particles, making them move

light An energy pathway by which energy is transferred from one store to another store

light-dependent resistor (LDR) A component whose resistance will decrease with increasing light intensity

light-emitting diode (LED) A component that emits light when a small current flows through it

loudness A measure of the intensity of a sound – in other words, how much energy the sound has

luminous A light source that produces and gives off its own light

newton The unit used to describe amount of force

opaque Not allowing light to pass through

parallel circuits A circuit that contains junctions where the electrical pathway divides or branches

pitch How high or low a sound is – affected by the frequency of the sound

poles The different ends of an electrical component – one will be positive and the other will be negative; or the ends of a magnet – north-seeking and south-seeking

refraction The bending of light or sound waves when it moves from one medium to another of different density

repulsion Pushing apart: for example, like (similar) poles of a magnet repel each other

resistor A component that is designed to reduce the current in a circuit

series circuit An electrical circuit with all of the components joined in a single loop

shadow An area behind an opaque object opposite a light source

sound wave A pattern of vibrations carried by sound through the air

thermal store of energy Giving an object more of this makes it hotter

translucent Allowing light to pass through but creating a change in the light rays so that the image is unclear

transparent Allowing light to pass through without changing the light rays, so that the image is clear

vacuum A space containing no air particles

vibration A pattern of movement of an object backward and forward, usually at high speed – there is no sound without vibration

weight The force of gravity that pulls an object towards another object (usually towards the Earth)